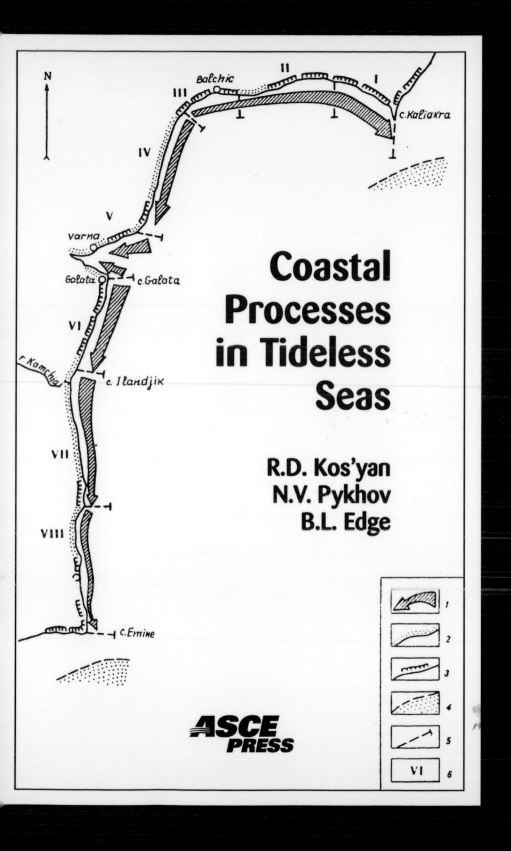

Coastal Processes in Tideless Seas

R.D. Kos'yan
N.V. Pykhov
B.L. Edge

ASCE PRESS

Coastal Processes in Tideless Seas

R.D. Kos'yan
N.V. Pykhov
B.L. Edge

PRESS
1801 Alexander Bell Drive
Reston, Virginia 20191-4400

Abstract: Originally published in 1991 by the Russian Academy of Sciences, Shirshov Institute of Oceanology, the authors have updated and translated this text into English. The book describes the modern state of the problem of sediment transport under wave influence in the coastal zone of tideless seas. Divided into three parts, this text addresses: hydrodynamic and elementary processes of sediment transport, characterization of research sites, measuring procedures, nearshore sediment mass transport, and some aspects of modeling.

Library of Congress Cataloging-in-Publication Data

Kos'ian, R. D. (Ruben Derenikovich)
[Gidrogennye peremeshseniia osadkov v beregovoi zone moria. English]
Coastal processes in tideless seas / R.D. Kos'ian, N.V. Pykhov, B.L. Edge.
p.   cm.
Includes bibliographical references and index.
ISBN 0-7844-0018-0
1. Coast changes. 2. Water waves. 3. Sediment transport. I. Pykhov, N. V. (Nikolai Valentinovich) II. Edge, Billy L.

GB451.2 .K6813   2000
551.45'7—dc21

00-023204

Contents

PART III
Sediment Transport   207

CHAPTER 7
Longshore Sediment Transport   209

CHAPTER 8
Cross-Shore Sediment Transport and Variability of the
Underwater Slope Profile   235

CHAPTER 9
Conclusions   263

References   267

Appendix   295

Index   301

Acknowledgments

The authors are grateful to S.M. Antsyferov, T. Basinski, V. Dachev, S.A. Efremov, S. Keremetchiev, B. Kirilova, A.D. Kochergin, O.L. Kuznetsov, I.O. Leont'ev, G.G. Mineev, H. Nikolov, E.L. Onischenko, V.I. Pakhomov, A.P. Phylippov, and N.S. Speransky for considerable assistance in the marine studies. The authors thank E.I. Birina, L.P. Fursova, T.M. Podymova, O. Ju. Potapenko, and O.V. Zaguskina for assistance in book design; Joyce Hyden for her editorial assistance and untiring support; students Sean Kelley and Ty Wamsley for their reviews, and Thorndike Saville, Jr., for his helpful review.

The authors owe special thanks to Professor V.V. Longinov for his constant support, kindly discussions, and helpful advice at all stages of research. His image will remain in our perpetual memory. It is a pleasure to acknowledge the ongoing support and encouragement of Mr. Orville Magoon for his assistance in the publishing of this book to whom we proudly dedicate this work.

Preface

The nearshore zone, though constituting only a small part of seas and oceans, plays an important role in human existence. Situated adjacent to the shore, it has been actively developed by people living in the narrow coastal zone. The first studies of nearshore dynamics were undertaken because of the concern of sailors over sea waves, which are especially dangerous in the vicinity of the coast.

Coasts without natural harbors of refuge had to be modified by hydraulic works (jetties, breakwaters, etc.) to protect vessels from storms. Works of this type were constructed in the Mediterranean by the Phoenicians, Greeks, and Romans as early as 500 B.C. (Edge, Magoon, and Baird, 1993). As ship drafts increased, the problem of silting arose, both for natural harbors in river mouths and for newly constructed ones. This problem could not be solved without the knowledge of sediment transport by waves and currents in the nearshore zone. In the second half of the twentieth century, rapid tonnage growth required the construction of navigation channels; prediction of silting in these channels became one of the important practical tasks of sediment transport studies. The solution to many present-day practical tasks depends on knowledge of the rate of sediment deposition and transport processes in the nearshore zone. Intake of sea water with minimal impurities for cooling systems of nuclear and thermal power stations is of vital importance, as the majority of these stations are being constructed on the coasts. There are some areas where quarrying of building materials (sand, shell) presents a serious problem since their removal from the seabed and beach in scientifically unreasonable quantities can result in severe ecological damage. Prediction of sediment discharge, deformation of the bottom relief, and change in the beach profile is vital for the security of various communications (cables, pipelines) and coastal structures (piers, unloading terminals, bridges), for support of recreational beaches, and for artificial beach creation.

Data on the frequency of sediment suspension, depth distribution, and transport modes are especially necessary for the solutions to ecological problems and development of mariculture, the importance of which has grown considerably. Considerable attention is being given to water and sediment flow, especially in the nearshore zone. In the "JGOFS" international project, one of the tasks being undertaken is the evaluation of seaward sediment discharge. The above examples show the significance of the nearshore zone in terms of human existence, though many other practical tasks and scientific problems could be listed.

Dominance of sediment transport caused by waves and various wave-induced currents is typical in the field of nearshore dynamics. During aqueous movement, sediment suspended in water acts as a passive agent, and its effect on the velocity field is generally insignificant. The intensity and direction of sediment transport depend on the surface wave spectrum, as well as the bottom relief and bottom sediment composition.

There are several approaches to describing the "nearshore zone" (Longinov, 1963; Longinov and Pykhov, 1981); the most widely recognized is defined as the zone where waves and wave currents actively affect the seabed. The line of maximal storm uprush is considered to be the shoreward limit of the nearshore zone, while its seaward limit is represented by either the depth at which wave motion affects the seabed (this depth is approximately half of typical wave length) or the depth at which bottom sediment motion commences, the depth of closure. These limits are dynamically mobile, thus the wave movement of ground swell can affect the seabed at the shelf edge, which could be considered as the nearshore zone. Hence, the term "nearshore zone" will be defined to include the sea area adjacent to shore and characterized by complete dissipation of wave energy. From this point of view, it can be divided into two parts separated by the wave breaking zone* and differing in the intensity of lithodynamic processes.

The area between the point of initiation of bottom sediment motion and the wave breaking point constitutes the outer part of the nearshore zone. It is characterized by the growth of wave height during its travel to shallower depths. Sediment motion is generally confined to the bottom boundary layer in this zone. The inner part of the nearshore zone is occupied by the surf zone, where active dissipation of wave energy occurs after wave breaking and secondary water motion begins (circulation, longshore currents, low-frequency waves) with a time scale greater than the period of typical waves. Together with surface waves, this secondary water motion controls bottom erosion and sediment suspension, transport, and deposition.

Although considerable experience in sediment transport studies has been accumulated, reliable predictions for particular nearshore environments is

---

*Terminology of the main elements of the nearshore zone is given in Chapter 1.

still impossible. Numerical models and laboratory experiments allow the exclusion of various factors which apply in natural environments, and are especially useful for understanding these processes in physical terms. Thus, modeling enables one to study processes in the wave bottom boundary layer, in which in-situ measurements are impossible to obtain with any confidence. On the other hand, modeling of irregular waves, water circulation, and sediment transport in time scales greater than wave periods is hardly possible under laboratory conditions. However, in natural environments, low-frequency waves are generated by irregular waves and are one of the main factors controlling nearshore circulation and the concentration field of suspended sediment. Because of this, considerable attention is now being directed toward in-situ sediment transport studies, the results of which form the basis for nearshore processes modeling. The most intensive studies of these processes began in the early 1970s, with the development of extensive international and national programs:

- JONSWAP (Joint North Sea Wave Project) (Hasselman, Barnett, and Bouws, 1973). Aspects of parameterization of wind waves spectrum were considered.
- EKAM-73 (Hupfer, Druet, and Kuznetsov, 1974). Ocean and atmospheric interaction was studied in the nearshore zone.
- LUBYATOWO-74,76 (Basinski, 1980), KAMCHIA,77-79 (Belberov and Antsyferov, 1985), and SHKORPILOVTZY-82,83,85,88. Experimental studies of hydrogenous sediment transport were carried out by CMEA countries.
- NSTS (Nearshore Sediment Transport Study), USA (Gable, 1981). Aimed at experimental data for modeling of nearshore sediment transport.
- CCSS (Canadian Coastal Sediment Study), Canada (Gillie, 1985). Investigation of hydrodynamics and sediment motion in the nearshore zone.
- NERC (Nearshore Environmental Research Center Study), Japan. Aimed at a study of nearshore processes for coastal geomorphic predictions.

Though embodying some specific features, these programs were all based on the same methodological concept: to construct physically valid models permitting the calculation of the sediment distribution and deposition in the nearshore zone using the original wind field, the surface wave spectrum, or the velocity field, respectively. These models also attempt to predict the bottom relief and shoreline modifications with sediment deposition variations. Despite intensive studies by the above programs, this idea is still far from realization. In this book the authors show the present state of the problem of examining the nearshore transport of sandy-silty sediments, using original data from in-situ studies along with those published by other authors. To illustrate the physical pattern of these processes, results of laboratory experiments and numerical modeling are widely employed. Discussion is confined to consideration of physical patterns and likelihood of sediment transport, for prediction purposes. Problems of underwater slope relief and shoreline and beach dynamics controlled by nearshore sediment mass movement are not discussed here and the reader is referred to the book *Nearshore Dynamics and Coastal Processes* (Horikawa, 1988).

This book is in three parts. Part one deals with present-day knowledge of hydrodynamic processes and elementary processes of sediment transport. In the second part, in-situ studies of coastal processes are discussed, research sites are characterized, measuring procedures are described, and the main results of bed microform and suspended sediment studies are given. Nearshore sediment mass transport and some aspects of its modeling are presented in the concluding third part.

Lastly, before moving into the detailed discussion in this volume, a word is in order about the title, *Coastal Processes in Tideless Seas.* First and foremost, the subject of dynamic sediment processes from rivers to coasts is very complex. The number of variables is quite large and represents several physical processes taking place simultaneously. These processes range from suspension of material to rounding of sediment particles to multiple turbulent scales. Of course, large-ranging tides—as is the case with many coastlines around the world—tend to smear the effect of these processes over a wide area on the active beach face and underwater as well. By concentrating only on those environments where the tidal range is small or nonexistent, the problem is greatly simplified, and the analysis of the data from field experiments is made more straightforward.

PART

I

Modern Concepts of Nearshore Hydrodynamics and Elementary Coastal Processes

CHAPTER 1

Nearshore Hydrodynamics

Nearshore water motion results from external forces in the form of tidal currents, swell and wind waves, various wave currents generated in the surf zone, storm surge, and tsunamis. In natural environments, these phenomena have various degrees of repeatability and differ in their effect on the bottom slope. Tsunami and high storm surge levels affect only few coasts of the World Ocean and even then they are not frequent. In most cases wind waves and swell control nearshore dynamic processes. The nearshore velocity field directly affects the bottom sediment. Energy dissipation, interaction, and breaking of the surface waves generate secondary motions such as long-period waves, longshore currents, and horizontal and vertical circulation. Spatial and temporal scales of these motions evidently depend on the scales of morphodynamic elements of the bottom slope and the shoreline.

The main energy source of wave formation is wind, constantly blowing above the water surface. Wave parameters depend on the wind velocity, duration, and fetch value. Wind waves near the generating area are usually irregular. When the wind dies out, or when waves leave the generating zone, they propagate as swell characterized by nearly constant period and form.

When waves propagate shoreward, they reach a certain point where their length becomes less than the local depth. From that moment, waves induce oscillatory motion of the near-bottom water, the amplitude of which increases with propagation into shallower depths. Under certain conditions, the movement and transport of bottom sediment begin. Waves are refracted, which results in the wave crests becoming parallel to isobaths. Simultaneously with refraction, wave deformation occurs—wave forms and heights are changed. With further deformation, the wave breaks and in the process of energy dissipation, water movements of various scales are induced, such as small-scale turbulence, large-scale vortices in a breaking wave, long-period waves, and nearshore currents. These features, created by the broken wave, determine the intensity and direction of sediment transport in the surf zone.

Seaward of the wave breaking zone, sediment motion is induced both by surface waves and by various currents (tidal, drift, etc.). In areas with tidal current velocities up to tens of centimeters per second, such currents generally control the resulting bottom sediment transport in the offshore zone (Soulsby, 1983). The superposition of surface waves and current intensifies near-bottom water motion and as a result, movement of transported sediments is also increased. Outside the wave breaking zone, sediment motion occurs close to the bottom, and for sediment transport modeling one should know the dynamics of the bottom boundary layer both for pure wave movement and for combined wave and current action.

This chapter presents the main characteristics of bottom boundary layers formed by surface waves, velocity field and turbulence in a breaking wave, long-period waves, and nearshore water circulation. These are the primary hydrodynamic factors that determine the physical characteristics of sediment transport discussed in this book. The applicability of various theories of waves, refraction, deformation, and statistics of a shallow-water wave field are discussed in numerous publications and therefore are not presented herein.

Prior to consideration of certain hydrodynamic processes, it is important to review some terms used in the description of nearshore hydrodynamics. Fig. 1 shows schematically the zone divisions and terms that are common in world-wide studies. The main terminology of the elements of the nearshore zone are:

Nearshore zone: The zone extending seaward from the landward limit of storm overwash to the limit of initiation of sediment movement.

Offshore zone: The zone from the breaker zone to the sea.

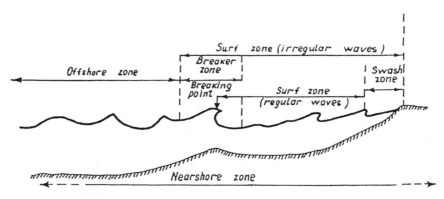
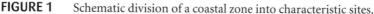

FIGURE 1   Schematic division of a coastal zone into characteristic sites.

Breaking zone: The zone of breaking of irregular waves.

Surf zone: The zone from the seaward limit of the breaker zone to the landward limit of the swash zone.

Breaking point: The starting point of wave breaking.

Swash zone: The zone from the wave uprush limit to the area of collision between backrushing water with incoming waves.

1.1 Bottom boundary layer of wave flow

When surface waves enter shallow water, oscillatory water movements create a boundary layer through friction against the bottom. This layer will be called the "oscillatory boundary layer" (OBL). Study of publications on shelf sediment transport shows that the OBL has recently received greatly increased attention. First, the OBL is the exact site of initiation of sediment transport, bottom erosion, and formation of microforms, which determine the essential boundary conditions for wave-induced sediment transport modeling. Second, the last two decades have witnessed intensive development of computation methods of turbulent currents. Having reduced the cost of calculations, computerization has also given a real opportunity for numerical modeling of the OBL. Third, development of the new generation of measurement devices, data collection, and processing systems enabled us to carry out a number of precise boundary layer measurements under laboratory conditions against which the suggested models were tested. It should be noted here that in situ measurements in the wave bottom boundary layer are still lacking as the available measuring devices do not permit measurements within a few-centimeter thick layer above the sea bottom.

Publication of the results from one of the two experiments, during which precise measurements in a turbulent current boundary layer were made (Johnsson, 1963, 1967) and a brief description of another experiment by Carlsen (Carlsen, 1967) marked a significant point in OBL studies. Complete and detailed descriptions of experimental results were published in 1976 (Johnsson and Carlsen, 1976). The importance and value of these experimental studies are proved by the fact that up until 1987 they were the only results throughout the world used for modeling velocity profiles, bottom shear stress, and other parameters of a turbulent OBL. Elegant results of Sleath's and Summer's experimental studies were published in 1987 (Sleath, 1987; Summer, Jensen and Fredsøe, 1987), which significantly widened the knowledge of transitional and turbulent regimes of an oscillatory boundary layer.

Kajiura (1968) suggested the first physically valid model of a turbulent OBL based on fluid mechanics achievements of that period. For model calibration, Johnsson and Carlsen's (1976) experimental results were used.

The aforementioned publications, along with giving some insight into the velocity field structure in an OBL, also presented numerous questions and formed a basis for further studies, the most interesting and significant of which were published in the 1980s (Grant and Madsen, 1979; Smith, 1977; Johnsson, 1980; Davies, 1986a,b; Myrhaugh, 1982; Sert, 1982; Fredsøe, 1984; Hagatun and Eidsvik, 1986; Trowbridge and Madsen, 1984; Sleath, 1987; Summer, Jensen, and Fredsøe, 1987; Justesen, 1987,1988; Davies, Soulsby, and King, 1988; Zheleznyak, 1988). The data presented in these publications form the basis for our current understanding of OBLs as documented in Fredsøe and Deigaard (1992). In the following sections, some aspects of OBL dynamics, most important in terms of sediment transport, are presented.

1.1.1 OBL determination and regimes of motion

Let us examine a flow of incompressible fluid above a flat rough bottom. Assume that movement is periodic and parallel to the bottom. In this case, the velocity of water at an infinite distance from the bottom can be written as:

$$U_\infty(t) = U_m \sin \omega t \qquad (1.1)$$

where U_∞ is the mean velocity at instant t at some distance from the bottom, U_m is its amplitude, and ω is angular frequency. Due to bottom friction, a velocity profile is formed, the typical form of which is shown in Fig. 2 for $\omega t = \pi/2$. Various OBL height determinations can be found in the literature. Johnsson (1980) suggests to set the OBL height (δ_w) equal to a minimal distance from bottom to a point where $U = U_m$. In other phases δ_w boundary layer height will be less than δ_m. Justesen in his work (1988) sets the OBL height (δ_m) equal to a minimal distance from the bottom to a point where

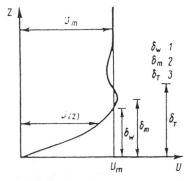

FIGURE 2   Determination of the principal elements of wave flow bottom boundary layer.

$\partial U/\partial z = 0$ (Fig. 2). Davies (1986a) introduces a height definition (δ_T) as a distance from the bottom to a point where the velocity difference from $U_\infty = U_m$ is 1% (Fig. 2). The same work determines boundary layer height not only by velocity profile, but by the vertical profile of turbulent energy. In that case, the boundary layer height (δ_E) is a minimal distance from the bottom to a point where turbulent energy is 1% of its maximal value at the bottom. The length scale is designated as $\delta_s = U_{*m}/\omega$, which makes the comparison of all the above definitions possible. (Here U_{*m} is the friction velocity amplitude.)

Near-bottom flow is typically characterized by the Reynolds amplitude number:

$$\mathrm{Re} = U_m a_m / \nu \tag{1.2}$$

or by the Reynolds number for bottom roughness

$$\mathrm{Re}_* = U_{*_m} K_s / \nu \tag{1.3}$$

where $a_m = U_m/\omega$ is the amplitude of water particle motion at the bottom, K_s is the linear size of bottom roughness elements, and ν is the kinematic viscosity. Water flow in a boundary layer can be laminar, turbulent, or transitional between these two. In laminar flow, motion in an OBL is determined by viscous forces.

According to Johnsson (1980), who studied and analyzed all the data on oscillatory flow regimes along smooth and rough surfaces, if the bottom is smooth, a laminar pattern for smooth walls exists at $\mathrm{Re} < 10^4$, while a turbulent boundary layer will exist at $\mathrm{Re} > 3 * 10^5$. When $10^4 < \mathrm{Re} < 3 * 10^5$, flow is considered to be transitional from laminar to turbulent. In this pattern, the flow character is determined both by viscous forces and by Reynolds stresses. For a rough bottom, the OBL motion pattern is determined not only by Re, but by the relative bottom roughness a_m/K_s.

With $\mathrm{Re}_* < 5$ the motion pattern is defined as hydrodynamically smooth. This condition is satisfied for weak bottom currents and for relatively smooth surfaces. For marine environments with $\nu = 1.2 \times 10^{-6}$ m²/s and $U_{*m} = 0.01$ m/s, this pattern will hold at $K_s = 0.006$ m, i.e., if a flat, even bottom is composed of medium-grained sand or by finer particles. In this case, a viscous sublayer directly adjoins the bottom.

At $\mathrm{Re}_* > 70$, the motion pattern is defined as hydrodynamically rough. In this flow condition, bottom roughness elements are larger than the viscous sublayer thickness, and oscillatory flow motion and its properties depend not on the Reynolds number, but on a_m/K_s. For a flat, sandy bottom composed of uniform grains, K_s is equal to their diameter d. For a flat bottom composed of heterogeneous material, Kamphuis (1975) experimentally obtained $K_s = 2d_{90}$, where d_{90} is the particle size with 90% of the particles having a diameter less than d_{90}. For a "not particularly smooth bottom," Nielsen (1979) suggests $K_s = 2d$, where d is the bottom sediment mean diameter. In actual marine conditions, the bottom is rarely smooth, with various scale micro- and macroforms present. In this case, the equivalent roughness K_s will be determined by form types and their characteristics since there are no appropriate methods of K_s determination. We list below some of the most frequently used estimates for two-dimensional ripples. In a number of studies (Johnsson, 1967; Tunstall and Inman, 1975; Hsiao and Shemdin, 1978) K_s is taken equal to 4 H_r (H_r is the ripple height). Swart (1976) gives $K_s = 25H_r^2/\lambda_r$ where λ_r is the ripple length. A similar equation with a coefficient value of 27.2 was obtained by Grant and Madsen (1982), who estimated bottom roughness in the presence of bed microforms and sediment transport on the basis of laboratory data analysis.

On the basis of experimental data, Johnsson (1980) suggested that for practical purposes a lower limit for the turbulent, hydrodynamically rough regime could be estimated by:

$$\text{Re} = 10^4 \quad \text{for} \quad 1 \le \left(\frac{a_m}{K_s} \right) \le 10 \tag{1.4}$$

$$\text{Re} = 10^3 \left(\frac{a_m}{K_s} \right) \quad \text{for} \quad 10 < \left(\frac{a_m}{K_s} \right) \le 10^3 \tag{1.5}$$

The height of the OBL is an important parameter. For the hydrodynamically smooth turbulent regime, after Johnsson

$$\frac{\delta_w}{a_m} = \frac{0.0465}{\sqrt[10]{\text{Re}}} \tag{1.6}$$

For the hydrodynamically rough regime

$$\frac{30\delta_w}{K_s} \log \frac{30\delta_w}{K_s} = 1.2 \frac{a_m}{K_s} \tag{1.7}$$

Estimates show that at $10 < (a_m/K_s) < 5 \times 10^2$ the boundary layer height is 2–4% of a_m.

For the laminar boundary layer, a height value is found from an analytical solution and is given by the equation:

$$\delta_v = \pi\sqrt{v/2\omega} \tag{1.8}$$

One of the important boundary layer characteristics is bottom shear stress, since it is a main parameter in describing sediment transport. For practical estimates, bottom stress (τ_B) is most frequently expressed through the bottom friction coefficient f_w and the velocity amplitude at the upper limit of the boundary layer:

$$\tau_B = \tfrac{1}{2}\rho f_w U_m^2 = \rho U_{*m}^2 \tag{1.9}$$

where ρ is water density and U_{*m} is the amplitude of the friction velocity. Empirical relationships for estimating f_w are also given by Johnsson (1980). For a hydrodynamically smooth turbulent regime:

$$\frac{1}{4\sqrt{f_w}} + \log\frac{1}{4\sqrt{f_w}} = \log \mathrm{Re} - 1.55 \tag{1.10}$$

For a hydrodynamically rough turbulent regime:

$$\frac{a_m}{K_s} > 1.57, \quad \frac{1}{4\sqrt{f_w}} + \log\frac{1}{4\sqrt{f_w}} = -0.08 + \log\left(\frac{a_m}{K_s}\right) \tag{1.11}$$

$$\frac{a_m}{K_s} \le 1.57, \quad f_w = 0.3 \tag{1.12}$$

Variation of f_w with a_m/K_s for hydrodynamically rough regime, from equations (1.11) and (1.12), by criteria suggested by other authors and their comparison with experimental data is shown in Fig. 3.

1.1.2 OBL modeling

Let two-dimensional water flow move above a horizontal bottom. With $2\pi\delta_w/\lambda \ll 1$, $2\pi a_m/\lambda \ll 1$ and $h/\delta_w \gg 1$ motion in the bottom boundary layer is described by the linear equation:

$$\frac{\partial U}{\partial t} = -\frac{1}{\rho}\frac{\partial P}{\partial x} + \frac{1}{\rho}\frac{\partial \tau}{\partial z} \tag{1.13}$$

where λ is the wave length on the sea surface, h is the local depth, τ is the shear stress, P is pressure, and t is time. The x-axis is oriented along the bot-

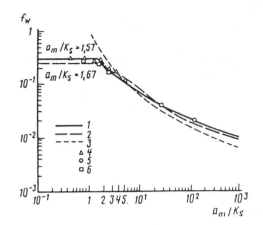

FIGURE 3   Variation of bottom friction coefficient in a wave field as a function of a_m/K_s (Johnsson, 1980). 1: semiempirical equation of Johnsson (1963); 2: Kajiura (1968); 3: empirical equation of Kamphuis (1975); 4: experimental data of Bagnold (1946); 5: experiment 1 of Johnsson (1963); 6: experiment 2 of Carlsen (1967).

tom, in the direction of wave propagation, and the z-axis is oriented upward from the bottom. Outside the boundary layer, fluid motion is described by the equation:

$$\frac{\partial U_\infty}{\partial t} = -\frac{1}{\rho}\frac{\partial P}{\partial x} \tag{1.14}$$

Substituting the pressure gradient $\partial P/\partial x$ from (1.14) and (1.13), we have:

$$\frac{\partial U}{\partial t} = \frac{\partial U_\infty}{\partial t} + \frac{1}{\rho}\frac{\partial \tau}{\partial z} \tag{1.15}$$

Stating velocity deficit as $U_g = U - U_\infty$ rewrite (1.15) in the form:

$$\frac{\partial U_g}{\partial t} = \frac{1}{\rho}\frac{\partial \tau}{\partial z} \tag{1.16}$$

The boundary conditions will be:

$$U(0,t) = 0 \text{ for a smooth bottom} \tag{1.17}$$

$$U(z_o,t) = 0 \text{ for a rough bottom} \tag{1.18}$$

$$U(z,t) = U_\infty \text{ for } z \Rightarrow \infty \qquad (1.19)$$

Water motion outside the boundary layer is periodic, $U_\infty(t) = U_m \sin \omega t$, where U_m is the velocity amplitude at the upper limit of the oscillatory boundary layer. In general, the shear stress τ from (1.15) consists of viscous and turbulent parts:

$$\tau = \rho\left(\nu\frac{\partial U}{\partial z} - \overline{U'W'}\right) \qquad (1.20)$$

where U' and W' are velocity fluctuation components along the x and z axes, respectively. Equation (1.15) can be integrated with respect to z:

$$\tau = r\int_z^\infty \frac{\partial}{\partial t}(U - U_\infty)dz \qquad (1.21)$$

Equation (1.21) is used for shear stress calculation by measuring the profile of $U(z)$. The turbulent component of the shear stress is usually expressed through turbulent viscosity:

$$-\overline{U'W'} = \nu_T\frac{\partial U}{\partial t} \qquad (1.22)$$

and

$$\tau = (\nu + \nu_T)\frac{\partial U}{\partial z} \qquad (1.23)$$

In the case of laminar flow $\overline{U'W'} = 0$, and the solution of equation (1.15) with boundary conditions (1.17)-(1.19) will be (Lamb, 1932):

$$U = U_m\left[\sin \omega t - e^{\beta z}\sin(\omega t - \beta z)\right] \qquad (1.24)$$

$$\tau = \sqrt{2\rho\nu\beta}U_m e^{-\beta z}\sin\left(\omega t - \beta z + \frac{\pi}{4}\right) \qquad (1.25)$$

where $1/\beta = \sqrt{2(\nu/\omega)}$. The velocity amplitude and shear stress variations with boundary layer height are shown in Fig. 4, which demonstrates the principal difference between laminar and turbulent boundary layers. Turbulent mixing, if compared to a laminar layer, results in rapid near-bottom velocity growth, while shear stress grows at some distance from the bottom.

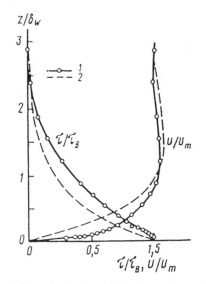

FIGURE 4 Vertical profiles of velocity amplitude and Reynolds stress in a wave bottom boundary layer (Johnsson, 1980). 1: turbulent boundary layer; 2: laminar boundary layer.

In the case of a turbulent boundary layer, to solve equation (1.15) one has to know either $\tau(z,t)$ or $v_T(z,t)$. All models of turbulent OBLs can be divided into three groups. The first group includes simple models in which the turbulent viscosity is a function of z and is time-invariant (Kajiura, 1968; Brevik, 1981; Myrhaugh, 1982; Aukrust and Brevik, 1985). The fundamental work of Kajiura (1968) was the first serious attempt to model an oscillatory boundary layer. By analogy with boundary layers of steady flows along smooth walls, the author used a three-layer v_T distribution in OBL height:

$$v_T = v \quad \text{for} \quad 0 < z < \delta_v \tag{1.26}$$

$$v_T = kU_{*m}z \quad \text{for} \quad \delta_v \leq z \leq z_1 \tag{1.27}$$

$$v_T = kU_{*m}z_1 \quad \text{for} \quad z_1 \leq z \leq \delta_w \tag{1.28}$$

where k is the Karman constant, v is the laminar sublayer thickness, and z_1 is the upper boundary layer, where the velocity distribution complies with the universal logarithmic law. Using the linear equation (1.15) with the boundary conditions (1.17)-(1.19), Kajiura obtained an analytical solution that was in good agreement with the experiments of Johnsson and Carlsen (1976).

For the case of a hydrodynamically rough regime of turbulent OBL motion, Myrhaugh (1982) suggested the parabolic law of v_T variation:

$$v_T = \frac{1}{2}\left[1-(\frac{z}{\delta_1}-1)^2\right] \quad \text{for} \quad \frac{K_s}{30} \leq z < \delta_i \tag{1.29}$$

$$v_T = \frac{1}{2}kU_{*m}\delta_i \quad \text{for} \quad \delta_i \leq z < \delta_w \tag{1.30}$$

where δ_i is the inner layer height, $U_{*m} = \sqrt{(\tau_B/\rho)}$ and τ_B is the bottom shear stress amplitude. In Fig. 5 the velocity and phase profiles relative to the velocity outside the boundary layer, calculated by this model, are compared with Johnsson and Carlsen (1976) experiments with $\omega t = \pi/2$. Relatively good agreement between the model and experimental data can be seen. This model also gives good predictions of velocity profiles for ωt (Fig. 6), which illustrate the vertical structure of an OBL velocity field.

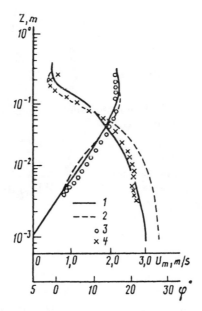

FIGURE 5   Comparison of velocity amplitude and phase, calculated from Myrhaugh's model (1982) with experiment 1 of Johnsson (1963). 1: calculated for $U_{*m} = 0.213$ m/s^{-1}, $\delta_i = 0.06$ m; 2: calculated for $U_{*m} = 0.213$ m/s^{-1}, $\delta_i = 0.03$ m; 3, 4: experimental data for velocity amplitude and phase, respectively (Myrhaugh, 1982).

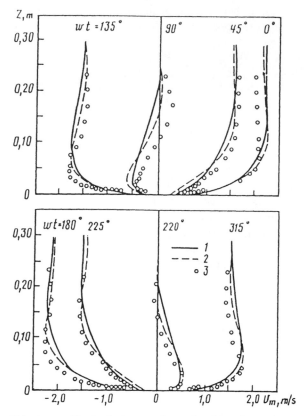

FIGURE 6   Velocity profiles in the bottom boundary layer for various ωt (Myrhaugh, 1982). 1, 2, 3 are as in Fig. 5.

Brevik (1981) also used a two-layer distribution of turbulent viscosity in his model, which in a later study was substituted by a three-layer distribution of ν_T (Aukrust and Brevik, 1985):

$$\nu_T = kU_{*m}z \quad \text{for} \quad \frac{K_s}{30} \le z < l_i \tag{1.31}$$

$$\nu_T = kU_{*m}[(nm-1)l_i - (m-1)z]/m(n-1) \quad \text{for} \quad l_i \le z \le nl_i \tag{1.32}$$

$$\nu_T = \frac{kU_{*m}l_i}{m} \quad \text{for} \quad z > nl_i \tag{1.33}$$

where l_i is the layer height in the universal law of velocity, fitted experimentally, and n and m are numerical constants. Results given in Fig. 7, as in above cases, are in good agreement with experimental data.

In the interval of $0.1\delta_w < z < 0.2\delta_w$, the velocity distribution is logarithmic. Good agreement is shown by all the models with the same experimental data, despite great variations of v_T, especially for the outer part of the boundary layer. This fact is difficult to explain reasonably in terms of the physics, as the form of v_T was chosen by the authors without sufficient physical justification and the resulting agreement was generally obtained by fitting of corresponding constant values. The linear equation (1.15) with the boundary conditions (1.17)-(1.19) cannot be sensitive to the form of v_T. Nevertheless and in spite of this, the above models, having analytical solution, simplify the calculations of velocity profile and bottom shear stress for a number of practical causes of sediment transport where great precision is not needed.

Compared with the first group of models, the second group is more accurate as the time variations of turbulent viscosity are introduced (Lavelle and Mofjield, 1983; Trowbridge and Madsen, 1984a,b; Davies, 1986a). To solve (1.15), Trowbridge and Madsen (1984a) took v_T in the form of

$$v_T = v_a \, \mathrm{Re}\left[1 + a_2(z)e^{i2\omega t}\right] \tag{1.34}$$

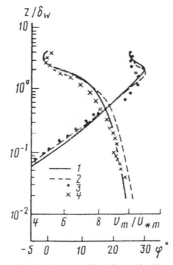

FIGURE 7   Comparison of velocity amplitude and phase, calculated from Aukrust and Brevik (1985) model with experiment 1 of Johnsson (1963). 1: calculated for $d_w = 6$ cm, $l_i = 2/3 \, d_w$, $m = 10$, $n = 3$; 2: calculated for $l_i = 1/2$ d_w, $m = 7.5$; $n = 4$; 3, 4: experimental data for velocity amplitude and phase (Aukrust and Brevik, 1985).

$$v_a = kU_* \begin{cases} z & \text{for} \quad \dfrac{K_s}{30} \le z \le \delta_i \\ \delta_i & \text{for} \quad z \ge \delta_i \end{cases} \tag{1.35}$$

where U_* is the mean period friction value, Re is the real part of the expression between the brackets, and $a_2(z)$ is the amplitude of the second harmonic, taken by the authors in the form of:

$$a_2(z) = 2e^{-i2\omega t}\left|\overline{U_*}\right| \tag{1.36}$$

The analytical solution is unwieldy and requires numerical methods of calculation. Comparison of theoretical calculations with the above experiments of Johnsson and Carlsen (1976) gives nearly the same agreement for velocity and phase profile, as for time-constant viscosity models (Fig. 8), but unlike the constant models, this model permits prediction of higher harmonics of velocity field and shear stress.

Davies (1986a) suggested an expression for $v_T(z,t)$, similar to (1.34) and (1.35), for a turbulent OBL in hydrodynamically rough flow:

$$v_T(z,t) = \frac{kU_{*m}}{1+\alpha} z(1 - \alpha\cos 2\omega t) \tag{1.37}$$

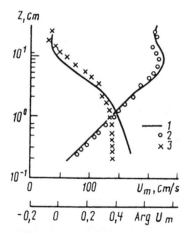

FIGURE 8   Comparison of velocity amplitude and phase: 1: calculated from the model of Trowbridge and Madsen (1984a); 2: with experiment 1 of Johnsson (1963); and 3: Trowbridge and Madsen, 1984b.

where α is numerical coefficient, varying from 0 to 1. $\alpha = 0$ corresponds to the case with time-steady-state and linear height variation of the turbulent viscosity. Examples of theoretical profiles for various α values and various phases ωt are shown in Fig. 9, clearly illustrating velocity profile dependence on this parameter. By comparison of his numerical calculations with that of other authors, Davies found $\alpha = 0.5$ to be a reasonable value for practical computations.

The models based on turbulent flow modeling, widely adopted for practical purposes in the last 20 years as a result of computerization (Johns, 1977; Justesen and Fredsøe, 1985; Asano and Iwagaki, 1986; Blondeux, 1987; Justesen, 1987, 1988; Audin and Shuto, 1988; Asano, Godo, and Iwagaki, 1988) come into the third group.

A detailed analysis of calculation methods for boundary layer turbulent currents is given by Lushik, Pavel'ev, and Yakubenko (1988). Of all the methods available for calculation of turbulent OBL parameters, only one- and two-parametric models are used now. As stated above, the equation of OBL motion (1.15) is not of closed form due to presence of $\tau = -\overline{U'W'}$ which should be defined also. Closure of the above models was attained by specifying the turbulent viscosity as a function of coordinates and time. With turbulent closure, the Prandtl hypothesis of mixing mode is frequently used, through which ν_T is expressed:

$$\nu_T = l^2 \left| \frac{\partial U}{\partial z} \right| \tag{1.38}$$

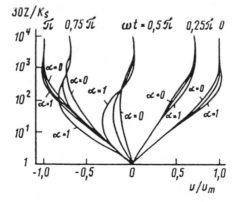

FIGURE 9   Theoretical profiles of velocity in various phases of wave flow (Davies, 1986a). $a_m/K_s = 159$, $\alpha = 0$, 0.5 and 1.0.

where l is the mixing length. Bakker (1974) and Johns (1975) used this approach although it has some limitations, as do those considered above, since v_T cannot be expressed directly through turbulence parameters and the solution does not account for turbulence development during the preceding half-period. This disadvantage is avoided in OBL models where the transfer equation for one of the turbulence parameters (turbulence kinetic energy, energy dissipation rate, turbulent viscosity, shearing stress) is used for closing of an equation (1.15). For modeling of the turbulent OBL, the transfer equation for turbulent kinetic energy is mostly used; it was suggested for a flat boundary layer by Kolmogorov as early as in 1942 (Kolmogorov, 1942):

$$\frac{dE}{dt} = \frac{\partial}{\partial z}\left(v_E \frac{\partial E}{\partial z}\right) + \tau \frac{\partial U}{\partial z} - \varepsilon_g \tag{1.39}$$

where $E = \frac{1}{2}(U'^2 + V'^2 + W'^2)$ is turbulent energy, ε_g is the rate of turbulent energy dissipation, v_E is a coefficient of turbulent energy diffusion defined as:

$$v_E = c_\mu \sqrt{E} \cdot L \tag{1.40}$$

where L is the turbulence integral scale, and c_μ is the proportionality coefficient. This approach has been implemented for the OBL by several researchers (Justesen and Fredsøe, 1985; Davies, 1986b; Justesen, 1987; Zheleznyak, 1988). The basic principles of these models are generally similar (differing only in boundary conditions, selection of constant numerical values and methods of numerical solution). The model formulated by Davies (1986b) is discussed below.

A flat turbulent OBL in a hydrodynamically rough regime is considered. As before, the OBL is assumed to be thin, i.e.:

$$\frac{2\pi\delta_w}{\lambda} \ll 1, \quad \frac{2\pi}{\lambda}a_m \ll 1, \quad \frac{\delta_w}{h} \ll 1$$

The basic equations of the model are:

$$\frac{\partial U}{\partial t} = \frac{\partial U_\infty}{\partial t} + \frac{\partial}{\partial z}\left(\frac{\tau}{\rho}\right) \tag{1.41}$$

$$\frac{\partial E}{\partial t} = v_T \left(\frac{\partial U}{\partial z}\right)^2 + \frac{\partial}{\partial z}\left(v_T \frac{\partial E}{\partial z}\right) - \varepsilon_g \tag{1.42}$$

$$\frac{\tau}{\rho} = \nu_T \frac{\partial U}{\partial z} \tag{1.43}$$

with the limiting conditions:

$$U = 0, \quad z = \frac{K_s}{30} \tag{1.44}$$

$$\nu_T \frac{\partial E}{\partial z} = 0, \quad z = \frac{K_s}{30}, \tag{1.45}$$

$$U = U_m \sin \omega t, \quad z \to \infty \tag{1.46}$$

$$\nu_T \frac{\partial E}{\partial z} = 0, \quad z \to \infty \tag{1.47}$$

The right part of equation (1.42) has three terms. The first one describes turbulent energy generation, the second describes its diffusion, and the third describes turbulent dissipation. The boundary conditions for the mean velocity U are similar to those of the above models. For turbulent kinetic energy (TKE), the energy flux is assumed to be lacking both at bottom and as $z \to \infty$.

Equations (1.41)-(1.47) are completed with the help of the A. Kolmogorov hypothesis:

$$\nu_T = c_0 L E^{\frac{1}{2}} \tag{1.48}$$

$$\varepsilon_g = c_1 \frac{E^{\frac{3}{2}}}{L} \tag{1.49}$$

where the length scale is given:

$$L = -k \frac{E^{\frac{1}{2}}}{L} \left[\frac{\partial}{\partial z} \left(\frac{E^{\frac{1}{2}}}{L} \right) \right]^{-1} \tag{1.50}$$

where k is the von Karman constant, c_0 and c_1 are constants. The length scale found from (1.50) satisfies the condition $L \to k \cdot z$ as $z \to (K_s/30)$, which is assumed for consideration of wall turbulence. Integration of (1.50) makes the equation more suitable for inclusion within the model:

$$L = -kE^{1/2} \left\{ \int_{K_s/30}^{z} E^{-1/2} dz + \frac{K_s}{30} E_0^{-1/2} \right\} \tag{1.51}$$

where E_o is the TKE at the $(K_s/30)$ level. For numerical calculations the following constants are used: $c_o = c^{1/4}$, $c_1 = c^{3/4}$, $c = 0.046$, $\alpha_1 = 0.7$, $k = 0.4$. Calculations were made for 26 vertical levels arranged in a logarithmic grid and for 60 time steps per period. The typical values of the main parameters, characterizing a hydrodynamically rough regime in marine conditions, were taken as: $T = 10$ s, $a_m/K_s = 500$, $U_m = 1$ m/s, $(K_s/30) = 0.01$. Fig. 10 shows the velocity profiles for a series of time steps of the first half of the sixth cycle (time steps 301-328). It is clearly seen that in the lower OBL $[z \cong (0.1$ to $0.2)\delta_s]$ the velocity profile is logarithmic most of the time, being the most extended at time step 316 ($\omega t = \pi/2$) and the least extended at time step 301 ($\omega t = 0$).

Vertical profiles of turbulent energy are shown on Fig. 11. Near the bottom, at $z < 0.1$ cm, turbulent energy grows with ωt reaching a maximum at time step 313 and then drops to a minimal value at step 328. Nearly all the profiles show maximal energy in the $10^{-1} < z \le 10$ cm interval. A stable energy distribution is typical for $10 < z \le 55$ cm. The zero energy level ($z = 55$ cm) lies outside the boundary layer. Temporal variations of bottom friction stress τ_B are shown in Fig. 12; it can be concluded that τ_B leads the velocity outside the boundary layer by approximately 15 degrees. Estimates of f_w and δ_w and their comparison with Johnsson and Carlsen (1976) and Kamphuis (1975)

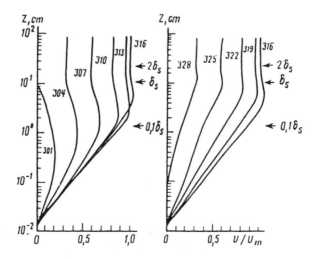

FIGURE 10   Calculated profiles of velocity for $U_m = 1$ m/s^{-1}, $T = 10$ s; $a_m/K_s = 530$ (Davies, 1986b).

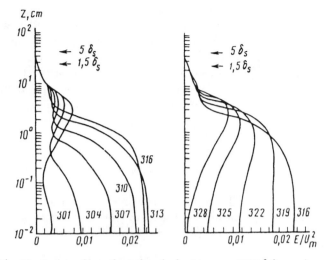

FIGURE 11 Vertical profiles of rated turbulent energy E/U_m^2 for various time steps. Initial parameters are the same as in Fig. 10.

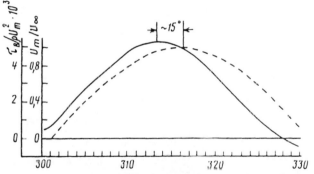

FIGURE 12 Temporal variations of velocity values U_m/U_∞ and Reynolds stress $\tau_B/\rho U_m^2$ for a wave half-period (time steps 301-330) (Davies, 1986b).

data showed this model to give satisfactory prediction of these parameters. If characteristic heights of the boundary layer are expressed through the length scale $\delta_s = U_m/\omega$, then $\delta_w = 1.3\delta_s$ and $\delta_E = 3\delta_s$, while the logarithmic layer height is $\delta_{\log} = 0.4\delta_s$ for $\omega t = \pi/2$. In general, this model provides for the calculation of both mean and turbulent characteristics of an OBL in hydrodynamically rough regime and with monochromatic waves.

The main limitation of the one-parameter models is the necessity for a length scale specification, which is not known in advance. In our case, this is equation (1.50), which is physically valid for currents. When complex currents are considered, such as those above bedforms large enough to be compared with the boundary layer height, or in the case of irregular waves, this assumption is far from evident, as turbulent flow characteristics at any given point and time will depend on flow prehistory and preset boundary conditions. Thus, if the L scale is not known in advance, two-parameter (so-called E-ε_g or k-ε) models are used. For this purpose, a transport equation for the rate of turbulent energy dissipation is added to the single-parameter model equations. Two-parameter models of hydrodynamically rough OBLs appeared recently in a number of studies (Hagatun and Eidsvik, 1986; Sato et al., 1986; Blondeux, 1987; Justesen, 1987, 1988; Zheleznyak, 1988); and models of OBLs with low Reynolds numbers, where fluid viscosity plays an important role, have been suggested by Japanese researchers (Asano, Godo, and Iwagaki, 1988; Audin and Shuto, 1988). The equations for a flat OBL above a hydrodynamically rough bottom has the form (Justesen, 1988):

$$\frac{\partial U}{\partial t} = \frac{\partial U_\infty}{\partial t} + \frac{\partial}{\partial z}\left(\nu_T \frac{\partial U}{\partial z}\right) \tag{1.52}$$

$$\frac{\partial E}{\partial t} = \frac{\partial}{\partial z}\left(\frac{\nu_T}{c_k} \frac{\partial E}{\partial z}\right) + \nu_T \left(\nu_T \frac{\partial U}{\partial z}\right)^2 - c\varepsilon_g \tag{1.53}$$

$$\frac{\partial \varepsilon_g}{\partial t} = \frac{\partial}{\partial z}\left(\frac{\nu_T}{c_g} \frac{\partial \varepsilon_g}{\partial z}\right) + c_1 \frac{\varepsilon_g}{E}\nu_T \left(\frac{\partial U}{\partial z}\right)^2 - c_0 c_2 \frac{\varepsilon_g^2}{E} \tag{1.54}$$

where $\varepsilon_g = E^{3/2}/L$, $E = \frac{1}{2}(U'^2 + W'^2)$, $U_\infty = U_m \sin \omega t$.
    Bottom boundary conditions are $z_o = (K_s/30)$:

$$U\left(\frac{K_s}{30}, t\right) = 0 \tag{1.55}$$

$$E\left(\frac{K_s}{30}, t\right) = \frac{1}{\sqrt{c_1}}\nu_T \left|\frac{\partial U}{\partial z}\right| \tag{1.56}$$

$$\varepsilon_g\left(\frac{K_s}{30}, t\right) = c_1^{3/4}\frac{E^{3/2}}{kz} \tag{1.57}$$

At the upper z_δ limit, formed by the upper level of the calculation grid

$$\frac{\partial U}{\partial z}(z_\delta,t) = 0 \tag{1.58}$$

$$\frac{\partial E}{\partial z}(z_\delta,t) = 0 \tag{1.59}$$

$$\frac{\partial \varepsilon_g}{\partial z}(z_\delta,t) = 0 \tag{1.60}$$

The bottom boundary condition in equation (1.56) is based on the assumed local equilibrium of generated and dissipated turbulent energy. The condition in equation (1.57) is taken in a form adopted for current boundary layers (Rodi, 1980). Computations typically employ the following constant values: $c_0 = 0.08$; $c_1 = 1.44$; $c_2 = 1.92$; $c_k = 1.0$; $c_g = 1.3$. It should be stressed that all the models for a hydrodynamically rough bottom are only valid at high values of a_m/K_s. An example of theoretical analysis of turbulent kinetic energy, length scale, turbulent viscosity, and shearing stress at $a_m/K_s - 10^3$ is shown in Fig. 13, which clearly demonstrates the alteration of these characteristics with boundary layer height and with time. The considerable vertical and temporal variations of turbulent kinetic energy (TKE) are impressive.

As one would expect, in line with the boundary conditions, TKE is maximal at the bottom. In the flow above, TKE levels depend on its diffusion from the near-bottom boundary layer. Mean and turbulent characteristics calculated by this model were compared with all available experimental data on hydrodynamically rough bottom (Johnsson and Carlsen, 1976; Sleath, 1987; Summer, Jensen, and Fredsøe, 1987) some results of which are given in Figs. 14 and 15.

The model gives a very good description of velocity profiles for different wave phases (Fig. 14). Calculated TKE profiles are in good agreement with experimental data. Difference is observed only at the bottom, where the experiment gives a lower TKE than that predicted by the model. In the model, the near-bottom TKE is determined by the boundary condition (1.56) obtained from the assumed local equilibrium between energy generation and dissipation, which is generally adopted for current flows. Fig. 15 shows that this assumption is not particularly true for the near-bottom oscillatory flow. According to Justesen (1988), the observed difference results from the fact that not all the phases of wave flow are characterized by a hydrodynamically rough regime. Here lies the possibility for model improvement, but neverthe-

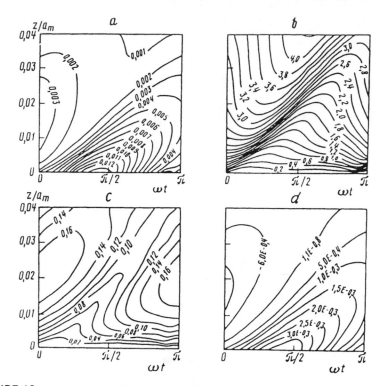

FIGURE 13 An example of calculation of turbulent kinetic energy E/U_m^2. 1: length scale, l/K_s; 2: turbulent viscosity, v_T/K_sU_m; and 3: Reynolds stress, $\tau_B/\rho U_m^2$ (d) in the z/a_m and ωt coordinates (Justesen, 1988).

less, physically acceptable suggestions on the boundary conditions of near-bottom energy are still lacking.

Until now, we have discussed OBL turbulent models for a hydrodynamically rough bottom. However, at low Reynolds numbers and over a smooth bottom, laminar to turbulent regimes can all occur during the same wave period, so the model should take account of molecular viscosity. This represents the real situations of bottom particle transport initiation and an early stage of bedform formation. In these cases the model construction should take account of the equal contribution of Reynolds stress and viscous stress to OBL dynamics. For wave boundary layers characterized by low Reynolds numbers, two-parameter models are generally used, in which the transport equation for turbulent energy and energy dissipation rate includes additional members characterizing energy losses due to near-bottom molecular viscosity (Audin and Shuto, 1988; Asano, Godo, and Iwagaki, 1988). For a flat OBL, the equations of the two-parameter model takes the form (Audin and Shuto, 1988):

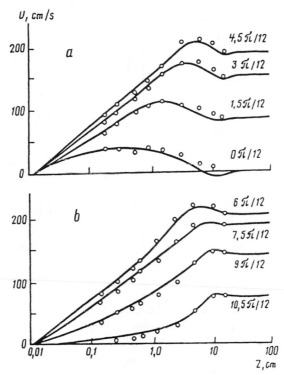

FIGURE 14 Comparison of velocity profiles calculated from the model of Justesen with the experimental data of Summer et al. (1987). 1: for acceleration and 2: retardation of wave flow, $a_m/K_s = 720$ (Justesen, 1988).

$$\frac{\partial U}{\partial t} = \frac{\partial U_\infty}{\partial t} + \frac{\partial}{\partial z}\left[(\nu + \nu_T)\frac{\partial U}{\partial z}\right] \qquad (1.61)$$

$$\frac{\partial E}{\partial t} = \frac{\partial}{\partial z}\left[\left(\nu + \frac{\nu_T}{c_E}\right)\frac{\partial E}{\partial z}\right] + (\nu_r + \nu_T)\left(\frac{\partial U}{\partial z}\right)^2 - \varepsilon_g - \frac{2\nu E}{z^2} \qquad (1.62)$$

$$\frac{\partial \varepsilon_g}{\partial t} = \frac{\partial}{\partial z}\left[\left(\nu + \frac{\nu_T}{c_g}\right)\frac{\partial \varepsilon_g}{\partial z}\right] - c_{\varepsilon 1}\frac{\varepsilon_g}{E}\nu_T\left(\frac{\partial U}{\partial z}\right)^2 - c_{\varepsilon 2}f_\varepsilon\frac{\varepsilon_g^2}{E} - f_{\varepsilon_g}\frac{2\nu\varepsilon_g}{z^2} \qquad (1.63)$$

where ν_r is some conventional viscosity dependent on bottom roughness. Comparison of these equations with (1.52)-(1.54) shows that the effect of near-bottom viscosity is allowed for with the help of terms $2\nu E/z^2$, $f_{\varepsilon_g}2\nu\varepsilon_g/z^2$

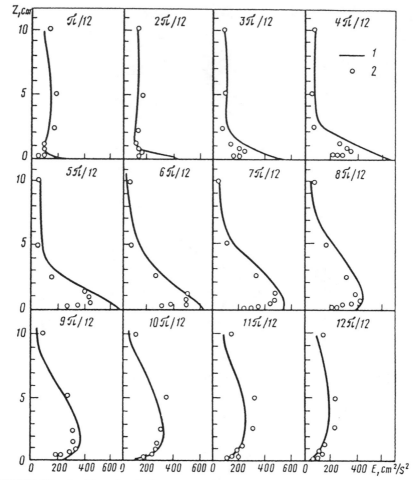

FIGURE 15   Profiles of turbulent energy for $a_m/K_s = 720$ (Justesen, 1988). 1: calculated from model; 2: experimental data of Summer, Jensen, and Fredsøe (1987).

and with term $v_r(\partial U/\partial z)^2$, which defines the additional generation of turbulent energy due to roughness. In (1.61)-(1.63) turbulent viscosity is defined as:

$$v_T = c_\mu f_m \frac{E^2}{\varepsilon_t} \tag{1.64}$$

and the rate of turbulent energy dissipation as:

$$\varepsilon_t = \varepsilon_g + \frac{2\nu E}{z^2} \tag{1.65}$$

The bottom boundary conditions at $z = 0$ are:

$$u = \varepsilon_g = E = 0 \tag{1.66}$$

at the upper boundary $z = \delta_w$

$$u = u_m \cos \omega t, \quad \frac{\partial E}{\partial z} = \frac{\partial \varepsilon_g}{\partial z} = 0 \tag{1.67}$$

Functions f_ε, $f_{\varepsilon g}$, f_m, and ν_r are selected empirically and their particular form is given in the work cited.

Fig. 16 shows variations of turbulent energy, turbulent energy production, and turbulent viscosity with OBL height and with time. A marked difference in turbulence during flow acceleration and deceleration phases is evident. Turbulent energy is dissipated completely at an early stage of flow acceleration which is approximately at $0.25 < t/T < 0.38$ ($T =$ wave period). With further flow acceleration, the bottom stress grows and generation of near-bottom turbulent energy begins again at $0.38 \leq t/T < 0.48$. At the flow deceleration phase, isolines of turbulent energy generation occupy the larger portion of the boundary layer leading to a discharge of turbulent kinetic energy toward the upper limit of the boundary layer. Greater turbulent energy is generated during flow deceleration than during acceleration.

Completing the analysis of readily available OBL models, we should make some specific observations. In this section we present some models differing in the extent of physical justification, and illustrate their possibilities. For practical tasks, selection of a model should depend on the requirements of the problem to be solved. For example, when energy losses by friction are analyzed for surface waves, or when bottom friction stress is evaluated for engineering calculations of sediment transport, simple models can be used that give rather good predictions of velocity profile and f_w as function of a_m/K_s. Single- and two-parameter models are excessive because of higher computational costs with an insignificant gain in accuracy.

For physical analysis of sediment transport by wave flow, the turbulent characteristics and the velocity field controlling the sediment motion should be known. Until now, turbulent characteristics were taken into account only by means of the turbulent viscosity as the key parameter in suspended sediment modeling. Thus, the one- and two-parameter OBL models, which give detailed information on spatial and temporal variations of turbulence energy,

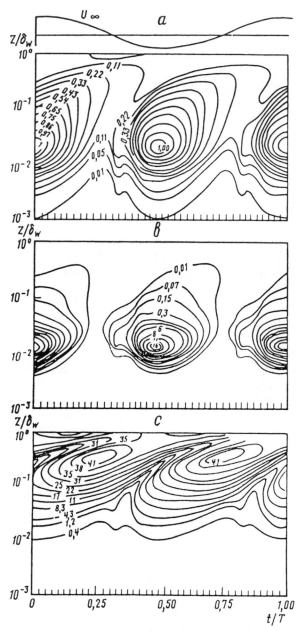

FIGURE 16 Variation of 1) turbulent energy, E/dU_m^2; 2) turbulent energy generation, $(\nu_T + \nu)(\partial U/\partial z)^2$; and 3) turbulent viscosity, $\nu_T/\nu(c)$ (Audin and Shuto, 1988).

its generation and dissipation, and on turbulent viscosity variations, closely studied here, are considered as most promising for the construction of physically valid numerical models of sediment suspension and transport. At present, computation based on these models is very time-consuming, but their common use for sediment transport modeling in engineering practice can be expected to be normal procedure in the near future. Nevertheless, there are some models of this type (Justesen and Fredsøe, 1985; Hagatun and Eidsvik, 1986) which will be considered in the next chapter in the discussion of suspended sediment.

1.1.3 Turbulence in the surf zone

During wave deformation, breaking, and evolution, the velocity field is controlled both by vortex-free (potential) and vortex flows formed after crest breaking, accompanied by wave energy dissipation (Battjes, 1988; Nadaoka, Hino, and Koyano, 1989). Let us consider a qualitative picture of vortex formation in a breaking wave. When waves are plunging, the wave crest is whirled along its propagation and a water jet falls forward of the crest into a trough, thus forming an air bubble, collapse of which produces a zone saturated with small air bubbles. When waves are broken by spilling, a similar process occurs locally, next to the top, and then the upper unstable portion of the crest runs off, forming a bore. Detailed descriptions of the wave breaking process can be found in Jansen's and Peregrine's studies (Jansen, 1986; Peregrine, 1987).

Fig. 17 shows the temporal evolution of the bubble saturated area and vortex structures forming in a breaking wave with the reference system migrating with the wave. Eddies with a horizontal axis whirl in the same direction, and in the areas of opposing velocities, strong velocity gradients appear that control a high degree of the energy dissipation of these coherent formations. During the destruction of these large-scale vortices, smaller ones appear, and starting from some moment they cannot be considered as coherent structures because of chaotic motion. The frontal motion of a breaking wave is also characterized by a scale reduction along with propagation, and from some moment a turbulent bore is formed. One of the first models of this is presented by Peregrine and Swendsen (1978).

Our present knowledge of the velocity field and turbulence of breaking waves is continuing to develop. Experimental studies carried out by Japanese researchers during the last 10 years under laboratory conditions with the help of modern measuring devices and flow visualization techniques (Nadaoka and Kondoh, 1982; Izumiya and Horikawa, 1982; Aono and Hattori, 1983; Aono, Ohashi, and Hattori, 1982; Hattori and Aono, 1985; Mizuguchi, 1986; Okayasu, Shibayama, and Mimura, 1987; Nadaoka, Hino, and Koyano, 1989)

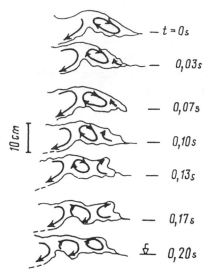

FIGURE 17   Temporal evolution of the bubble saturated area and vortex structures in spilling breaking wave with the reference system migrating at wave propagation velocity (Nadaoka, Hino, and Koyano, 1989).

played an important role in understanding breaking waves. When turbulence is analyzed in nonstationary and irregular flow, the selection of the turbulent component of velocity is of great significance. For that purpose, the methods of phase averaging, filtration, and coherent techniques are used now. According to the phase method, pulsation components are defined as deviation from mean values during a series of wave cycles for the given particular phase of wave motion (Nadaoka, Hino, and Koyano, 1989). This method is good only for regular waves and cannot be used for irregular waves. In accordance with the filtration technique, only those Fourier component, whose frequencies are higher than some given values are considered as turbulent (Nadaoka and Kondoh, 1982). In the coherent method, turbulent velocities are defined as incoherent motion with water surface fluctuations (Thornton, 1979; Aono and Hattori, 1983). This method is most frequently used in field conditions.

Nadaoka and Kondoh applied the filtration method for isolation of velocity pulsations to analyze velocity fields in the surf zone under laboratory conditions for a flat sloping bottom (Nadaoka and Kondoh, 1982). The velocity was measured by laser Doppler anemometer. Signal fluctuations passing through a filter at the 10 to 100 Hz frequency band were considered as turbulent. Special spectral analysis showed that this frequency band defines turbulence seaward of the plunge point and can cut off low-frequency components (large-scale eddies) around this zone. Fig. 18 shows the vertical profiles of the mean-square values of horizontal and vertical velocity components from two

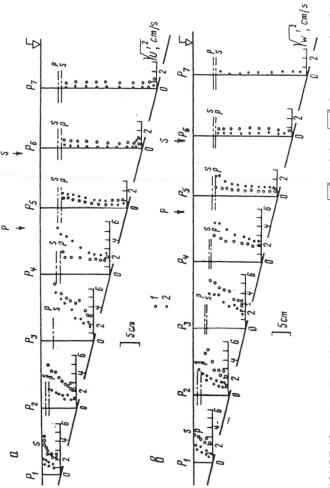
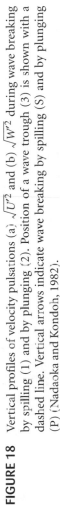

FIGURE 18   Vertical profiles of velocity pulsations (a) $\sqrt{U'^2}$ and (b) $\sqrt{W'^2}$ during wave breaking by spilling (1) and by plunging (2). Position of a wave trough (3) is shown with a dashed line. Vertical arrows indicate wave breaking by spilling (S) and by plunging (P) (Nadaoka and Kondoh, 1982).

experiments, differing in mode of wave breaking. Values of $\sqrt{U'^2}$ and $\sqrt{W'^2}$ seaward of the breaking point are maximal at the bottom, suggesting that the bottom boundary layer acts as a source of turbulence. After wave breaking, velocity fluctuations grow rapidly both at the bottom and upward, which is explained by the existence of vortex structures, their energy dissipation in areas of significant velocity shift between them, and penetration of turbulent motions toward the bottom. Accurate visual observations showed that the zone of strong velocity pulsations corresponded to the areas of high air bubble concentration (Nadaoka and Kondoh, 1982; Nadaoka, Hino, and Koyano, 1989). The bore height is the typical length scale. As a result of their visual and instrumental observations, Nadaoka et al. (1982, 1989) suggested a scheme of turbulence generation process in the breaking zone (Fig. 19) and a two-layer conceptual model of turbulence in the zone of wave deformation and breaking (Fig. 20). According to this model, the upper part of the water column is characterized by the existence of large-scale vortices with a horizontal axis parallel to the crest axis, definitely marked by an area of high bubble concentration. The destruction of these two-dimensional vortices on the back of the crest and intensive energy dissipation in between zones result in formation of three-dimensional vortices with a forward tilt of axis, sometimes reaching the bottom of the surf zone.

In close proximity to the bottom, turbulence is generated in the bottom boundary layer, to which small-scale turbulence penetrating from the upper layer is added. It was shown (Fig. 18), that turbulent pulsations of velocity are much lower at the bottom than in the water column, which is attributed to experimental bottom smoothness. In natural environments, where smooth bottoms do not exist, the intensity of near-bottom turbulence should be much higher. Seaward of the plunge point, turbulence is generated only in the bottom boundary layer.

Quantitative modeling of breaking wave turbulence presents a more difficult task. The basic equations, discussed in Section 1.1, are not suitable here and some new ideas of turbulent energy calculations are needed. If the vertical distribution of energy is known, one can evaluate the breaking wave turbulent viscosity profile, and from it proceed to the modeling of suspended sediment.

Battjes was one of the first who tried to construct such a model (Battjes, 1975). According to Battjes, the source of turbulent kinetic energy is the dissipation of wave energy in the surf zone during wave breaking. Assuming local equilibrium between turbulent energy generation and dissipation is the surf zone, he got $\overline{E}^{1/2} \approx (D/\rho)^{1/3}$ where $\overline{E}$ is the wave period and depth-averaged turbulent kinetic energy per unit mass and D is the wave energy dissipation per unit of bottom area. This relation was confirmed experimentally with an $\overline{E}$ value being over the measured width of the surf zone (Battjes, 1983). This idea was employed by Swendsen for analysis of the surf zone turbulence prob-

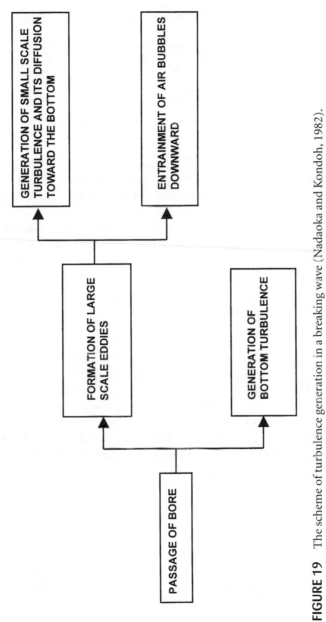

FIGURE 19   The scheme of turbulence generation in a breaking wave (Nadaoka and Kondoh, 1982).

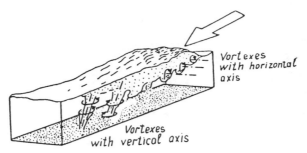

FIGURE 20   Large-scale vortex structures in the surf zone (Nadaoka, Hino, and Koyano, 1989).

lem (Swendsen, 1987). He assumed that the normalized value of turbulent energy $\overline{E}^{1/2} \approx (D/\rho)^{1/3}$ is only a function of the dimensionless height above the bottom (z/h). Swendsen tried to find the form of this function from experiments by Stive and Wind (Stive and Wind, 1982). The result is shown in Fig. 21. Despite scatter in the data, turbulent energy evidently tends to increase with z/h growth. Such variation is not unexpected, as the turbulence source occupies the surface and points to strong mixing.

The regime of bore progression and damping is defined by wave energy dissipation (D). Detailed discussion of the various approaches to determine the functional form of D in relation to the wave parameters is presented in recent publications on surf zone dynamics (Leont'ev, 1988; Battjes, 1988).

Due to laboratory experiments, the wave breaking and turbulence formation process is generally presented as a qualitative pattern, which serves as a physical basis for the modeling of sediment suspension and mixing in that zone. One can hardly hope that equally precise measurements of a breaking wave velocity field could be made in the near future in natural marine environments as the equipment is lacking which could endure high dynamic loads and afford measurements in irregular wave conditions with the water column saturated with sediment and air bubbles.

1.2 Horizontal water circulation

When wave deformation starts over the offshore slope, nearshore circulation at time scales greater than wave periods frequently occur. In a general case, it is three-dimensional and depends on the bottom relief and on the range of surface waves. Since the system of equations for a three-dimensional definition of water circulation is too complicated, circulation in a vertical plane normal to shore and in the horizontal plane is modeled separately. First, let us consider horizontal water circulation in the nearshore zone, the modeling of

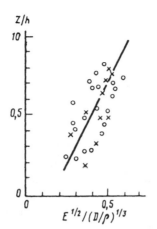

FIGURE 21   Variations of normalized values of turbulent energy $\bar{E}^{1/2} \approx (D/\rho)^{1/3}$ with distance from bottom. Crosses and circles indicate experimental data of Stive and Wind (1986). (Swendsen, 1987).

which has advanced significantly since Longuett-Higgins and Stewart (1960) introduced their radiation stress concept.

Horizontal circulation is defined by the equations of mass and momentum preservation, averaged over wave period and depth in approximation to plane current (Philips, 1977):

$$\frac{\partial(\rho\bar{\eta})}{\partial t} + \frac{\partial M_i}{\partial x_i} = 0 \qquad (1.68)$$

$$\frac{\partial M_i}{\partial t} + \frac{\partial}{\partial x_j}(U_j M_i + T_{ij} + T'_{ij}) + \rho g h \frac{\partial\bar{\eta}}{\partial x_i} + \tau_i = 0 \qquad (1.69)$$

where M is the mean value of mass per unit length or of momentum per unit area, $U = M/\rho h$ is the depth averaged velocity of the flow, and T_{ij} and T'_{ij} define (ij) components of radiation stress and depth averaged Reynolds stresses; $\bar{\eta}$ is mean water surface position relative to the calm water level, and τ_i is the mean friction stress at the bottom. Equations (1.68) and (1.69) allow for time variation on scales greater than the wave periods. Radiation stress gradients act as a principal driving force of circulation; in the general case they are written as:

$$-\frac{\partial T_{ij}}{\partial x_j} = h\frac{\partial}{\partial x_i}\left\{\frac{\omega Ec_g}{g^2}\frac{d(\tanh^{-1}kh)}{d(K_0h)}\right\} + \frac{D}{c_0}e_i \tag{1.70}$$

where Ec_g is the wave energy flow, E is the wave energy density, c_g and c_0 are the wave group and phase velocities, $K_0h = \omega^2h/g = kh$ tanh kh, k is a wave number, e_i is the unit vector components in wave motion direction, and D is the wave energy dissipation.

The first model of nearshore circulation based on the above general equations in approximation to a flat sloping bottom and the normal landfall of monochromatic waves with longshore periodic height variations, was suggested by Bowen (1969a). According to this model, longshore periodic height variation results in an uneven wave set-up distribution, the level of which is proportional to the breaking wave height. Because of the wave set-up difference, a pressure gradient develops and becomes a driving force of longshore currents directed from areas with maximal breaking wave heights to the places with minimal height. Currents observed in areas with minimal wave set-up are turned seaward and form a rip current. Farther from shore, deceleration of the rip current occurs. Behind the wave breaking line it widens and then returns shoreward, coming as a wide decelerated flow to areas of high wave breaking, and thus circulation cells are formed (Fig. 22). Under steady waves and uniform longshore conditions, these cells become cyclic, which is finally reflected in the cyclic development of relief forms (cusps, crescentic

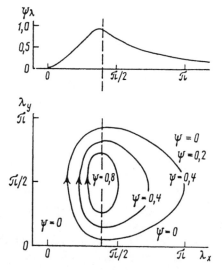

FIGURE 22   An example of a circulatory cell calculation (Bowen, 1969a). The wave breaking line is shown with a dashed line.

bars) which, in their turn, can affect further development of nearshore circulation. In principle, natural observations can hardly reveal the original factor of circulation—whether the existing bottom relief causes nonuniform wave set-up when waves of similar height are reaching shore, or nonuniformity in wave height leads to uneven longshore wave set-up, development of circulation, sediment transport and, finally, to given relief formation. Experiments in wave basins with a stiff, nonerodible bottom and normal monochromatic waves produce rhythmic pattern of circulatory cells, testifying that bottom topography is not necessarily the original reason for horizontal circulation. Edge waves, discussed below, are now considered to be one of the most probable reasons for the height modulation of wind-induced waves (Bowen, 1969a; Bowen and Inman, 1969; Sasaki and Horikawa, 1975). Field observations show that the periodic scale of rip currents (see Fig. 23) corresponds to the length of some edge wave modes (Bowen and Inman, 1969). On the other hand, the longshore bottom relief is always non-uniform in natural conditions and wave refraction leads to uneven longshore wave set-up even when waves have the same height. Sonu's results on a coast section with two wide shoals separated by a deep trough are considered as the classical example of circulation caused by bottom topography (Sonu, 1972). Under swell, two circulation cells were formed with slow shoreward currents above the shoals and a strong narrow seaward rip current in the trough, where mean water level was lower than in the shoals. In this situation wave landfall was nearly normal without significant longshore height modulation outside the breaking zone.

Complex bottom topography is usually described by numerical models. One of the first of such models was suggested for a few variants of underwater bottom topography (Noda, 1974). Later, with the development of physical

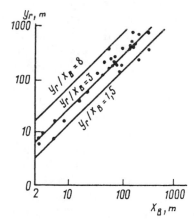

FIGURE 23   Correlation between surf zone width X_B and distance scale between rip currents Y_r (Sasaki and Horikawa, 1975).

knowledge and advanced numerical methods for solving the governing equations, a number of models were suggested for an arbitrary bottom relief (Wu and Liu, 1985; Wind and Vrengdenhill, 1986). We omit the description of particular models and note only that all of them are based on equations (1.68)-(1.70) and differ in their approach to the definition of bottom frictional stress, turbulent viscosity coefficient, and wave energy dissipation.

All the above models consider horizontal circulation for regular waves. Irregular waves are considered in the Leont'ev circulation model based on linearized equations of mass and momentum preservation (Leont'ev, 1987, 1988). He examines both variants of nearshore circulation generation—by periodic modulations of wave heights—and of the longshore bottom relief. Numerical evaluations show that in the case of periodic variations of bottom relief, current behavior depends on the relation between surf zone width (L_x) and longshore relief periodicity spacing (L_y). When L_x/L_y grows, circulation intensifies and becomes more asymmetric—seaward currents become stronger than shoreward ones. In this model, current asymmetry is determined by the irregularity of wind waves; here lies its principle difference from regular wave models, in which asymmetry of water motions is related to the preservation of convective transfer in equations of mass and momentum preservation.

Another factor causing nearshore circulation is hydrodynamic instability (Hino, 1974; Miller and Barcilon, 1978; Dalrymple and Lozano, 1978). Hino studied the stability of a normal wave system above a flat erodible bottom. He showed that this system is unstable toward infinitesimal longshore disturbances which increase greatly if their wave length is approximately four times greater than the surf zone width. The observed spacing of rip current periodicity as a function of surf zone width observed on USA, Mexican, South African, and Japanese coasts, indirectly confirms Hino's conclusions.

In later works (Dalrymple and Lozano, 1978; Miller and Barcilon, 1978), Hino's idea was developed for a non-erodible inclined bottom with due account of the inverse effect of wave-field currents induced by hydrodynamic instability. The periodic spacing of longshore circulation cells depends on the relationship between bottom slope and bottom friction coefficient and increases with the growth of this ratio.

In natural environments, unlike in models, the bottom configuration is very complex and contains three-dimensional forms. The situation is further complicated by existence of coastal structures (weirs, piers, groins, etc.) which exert a pronounced effect on water circulation. These are additional complications that arise when knowledge of the principal elements of circulation is required. Examples of water circulation at some Japanese coasts, measured by Lagrangian method, can be found in the Sasaki and Horikawa study (Sasaki and Horikawa, 1975) and in Horikawa (1988). Numerical modeling of environments with complex bottom relief, where refraction, diffraction, deformation, and wave breaking should be allowed for to estimate the turbulent parameters and circulation more accurately, is an important problem for the near future.

1.3 Longshore currents

When considering nearshore horizontal circulation, we noted that longshore currents could arise due to the level gradient produced by flow non-uniformity even at normal wave approach. Longshore currents of another type, called energetic (Shadrin, 1972), are generated by oblique wave approach due to the longshore flux of wave energy. Detailed analysis of numerous models designed for evaluation of longshore current velocities is given by Voitsekhovich (1986b) and Leont'ev (1988). Most physical models of such currents are based on equations (1.68)-(1.70) (Bowen, 1969a; Longuett-Higgins, 1970; Thornton, 1970). The most acknowledged model for flat inclined bottom and small angles of wave approach is that by Longuett-Higgins (1970). According to this model for uniform longshore conditions, the equation of momentum preservation, allowing for radiation stress, turbulent horizontal mixing, and bottom friction can be written in dimensionless form:

$$P\frac{d}{dx}\left[X^{5/2}\frac{dV^*}{dX}\right] - X^{1/2}V^* = \begin{cases} -X^{3/2}, & 0 < X < 1 \\ 0, & 1 < X < \infty \end{cases} \tag{1.71}$$

$$V^* = \frac{\upsilon}{\upsilon_B}, X = \frac{x}{x_B}, \upsilon_B = \frac{5\pi}{16}\frac{\gamma}{f_w}\sqrt{g(h+\overline{\eta})_B}\tan\beta\sin\alpha_B \tag{1.72}$$

where υ is time- and depth-averaged velocity of the longshore current, υ_B is the current velocity at the wave breaking line, x_B is the surf zone width, γ is the wave height to depth ratio, f_w is the bottom friction coefficient, α_B is the wave approach angle at the breaking line, $\tan\beta$ is the bottom slope, $P = \pi\tan\beta N/2kf_w$, $k = 0.41 = const$, and N is a dimensionless constant specifying the coefficient of horizontal turbulent mixing. The physical sense of the parameter N resides in that it reflects the role of horizontal turbulent mixing in the formation of longshore current regime. If the solution in the interval $0 < X < \infty$ is bounded and the condition of continuity is satisfied for V^* and dV^*/dX with $X = 1$, the solution of equation (1.71) takes the form:

for $P \neq 0.4$ $\qquad\qquad V^* = B_1 X^{P1} + AX \qquad 0 < X < 1 \tag{1.73}$

$$V^* = B_2 X^{P2} \qquad 1 < X < \infty \tag{1.74}$$

$$\text{for } P = 0.4 \qquad V^* = \begin{cases} \dfrac{10}{49}X - \dfrac{5}{7}X \ln X & 0 < X < 1 \\[3mm] \dfrac{10}{49}X^{-5/2} & 1 < X < \infty \end{cases} \qquad (1.75)$$

where

$$P_1 = -\frac{3}{4} + \left(\frac{9}{16} + \frac{1}{P}\right)^{1/2}, \quad P_2 = -\frac{3}{4} - \left(\frac{9}{16} + \frac{1}{P}\right)^{1/2}$$

$$B_1 = [P(1 - P_1)(P_1 - P_2)]^{-1}, \quad B_2 = [P(1 - P_2)(P_1 - P_2)]^{-1}, \quad A = \left[1 - \frac{5}{2}\right]^{-1} \quad (1.76)$$

Fig. 24 shows velocity profiles calculated for various P values. Without horizontal turbulent mixing ($N = 0$ and $P = 0$), the longshore current velocity grows linearly from the water edge to the breaking point. As mixing intensifies, the velocity maximum migrates shoreward and the current leaves the surf zone. This model gives rise to later models designed for longshore current calculations and allowing for: larger angles of wave approach (Kraus and Sasaki, 1979; Liu and Dalrymple, 1978), larger current velocities (Ostendorf and Madsen, 1979), a bottom profile other than rectilinear (McDougal and Hudspeth, 1983; Smith and Kraus, 1987), and wave field irregularity (Van De Graaf and Van Overeem, 1979; Thornton and Guza, 1986; Leont'ev, 1988). Some of the numerical models allow for real bottom topography but disre-

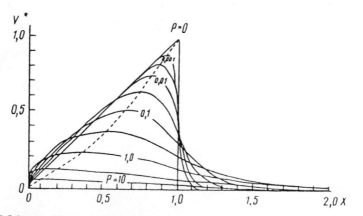

FIGURE 24   Profiles of longshore current velocity calculated for various P (Longuett-Higgins, 1970).

gard horizontal turbulent mixing (Sasaki and Horikawa, 1975; Birkemeier and Dalrymple, 1975). One of the most complete numerical models of nearshore currents, allowing for horizontal turbulent exchange, bottom friction, real bottom topography, and convective members in the equation for momentum transfer, was suggested by Wu and Liu (1985). This model was used for comparison of the calculated values with the results of in situ measurements of longshore current velocity profiles, obtained during NSTS experiments for an approximately flat bottom and uniform longshore conditions (Wu et al., 1985). The calculation results for $f_w = 0.01$ and $N = 0.014$ [the same values were used by Longuett-Higgins (1970)] are in good agreement with the measured velocity distribution (Fig. 25). The authors point out that the account of nonlinear members in the equations results in a lower maximal velocity if compared to the linear model, and gives a better approximation to field observations.

Wave irregularity is allowed for in Leont'ev's model (Leont'ev, 1988) which gives an analytical solution in approximation to low current velocities (relative to orbital) velocities, small angles of wave approach, and fair depth variation with positive values of bottom slope. Calculations by this model compared with NSTS experimental results (Fig. 25) showed good agreement. The fact that calculations made with the model suggested by American researchers and by the Leont'ev model, gave good results when compared with NSTS in situ measurements, evidently shows that the account of nonlinear members for regular waves has the same effect as does the account of irregular waves for the linear problem. The same fact was already noted when horizontal circulation was discussed.

All the above models define the velocity distribution over the surf zone width, characterized by a single maximum. In natural environments, two types of longshore current velocity are frequently observed. Such distributions were observed in the Kamchia testing site (Fig. 26) (Leont'ev and Pykhov, 1988). Smooth slopes on seaward profile sections and relatively steep slopes next to the water edge provided for the realization of either unimodal or bimodal velocity distribution. Under low swell, wave breaking occurred close to shore and the velocity distribution profile had only one peak. Under stronger swell, one peak remained near the water edge while another one, coinciding with the area of the highest wave breaking, appeared. A bimodal velocity distribution is not defined by available models, as at small slopes or along profiles with underwater bars, the surf zone can consist of isolated sites of wave energy dissipation separated by areas without wave breaking (Leont'ev, 1988). The character of the velocity distribution along such bottom profiles depends on the type of wave breaking and the underwater slope form. Fig. 27 schematically shows observed distributions of longshore current velocities for various profile types and various types of wave breaking derived by the generalization of numerous in situ measurements (Voitsekhovich, 1983).

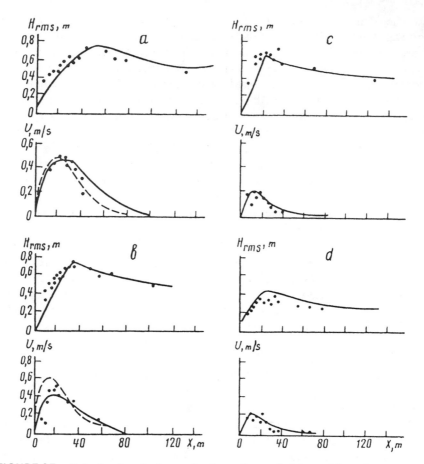

FIGURE 25   Calculated and observed values of wave height and longshore current velocity (Wu, Thornton, and Guza, 1985). a) H_o =0.52 m; $\overline{T}$ =14.2 s; αB =19.1°; f_w = 0.010; N = 0.01. b) H_o = 0.37 m; $\overline{T}$ = 16.0 s; αB = 15.2°; f_w = 0.015, N = 0.01. c) H_o = 0.50 m; $\overline{T}$ = 14.3 s; αB = 17.6°; N = 0.01. d) H_o = 0.26 m; $\overline{T}$ = 11.6 s; αB = 18.7°; f_w = 0.009; N = 0.01. Dashed line: calculations by linear theory; solid line: by nonlinear model; α_B: angle between wave beam and normal to the shore; H_{rms}: root-mean-square wave height; H_o: deep water wave height.

To define the bimodal curves of velocity profiles, Leont'ev (1988) suggested a model for a bottom profile with an underwater bar separated into three zones. The first zone corresponds to the seaward bar slope with the water depth decreasing monotonically so the above-mentioned model of this author is valid for this region. The second zone occupies an area around the bar top and is called a transitional area. As the water depth changes insignifi-

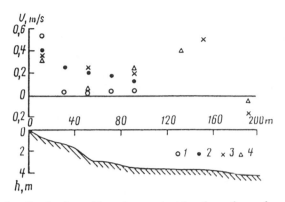

FIGURE 26   The distribution of longshore velocities along the underwater slope profile (Leont'ev and Pykhov, 1988). a) Series 4; $H_{rms} = 0.68$ m; $\overline{T} = 5.1$ s; $\alpha B = 30°$; b) Series 5; $H_{rms} = 1.23$ m; $\overline{T} = 5.7$ s; $\alpha B = 30°$; c) Series 6; $H_{rms} = 1.55$ m; $\overline{T} = 7.3$ s; $\alpha B - 30°$; d) Series 7; $H_{rms} = 1.41$ m; $\overline{T} = 8.3$ s; $\alpha B = 10°$.

cantly, it is considered to be approximately equal to the depth at the bar top. The third zone coincides with the shoreward bar slope. On the basis of force balance in these areas, Leont'ev derived the analytical equation of velocity distribution. Values calculated by them are in satisfactory agreement with the results of in situ measurements.

1.4 Vertical circulation

The definition of water motion by depth-averaged equations employed for the modeling of horizontal circulation and longshore currents is also suitable for the solution of problems involving calculations of nearshore sediment processes. For a better insight into processes of water and sediment transport normal to the shore, the vertical distribution of currents should be known. Let us consider the simplest case with monochromatic waves approaching shore at a normal angle and isobaths parallel to shore. The gradient of the wave setup produced by wave mass transport results in an undertow. The relation between these principal elements determines the form of the resulting transport velocity versus depth curve. For this simple case, the equation of momentum balance can be written in the form (Stive and Wind, 1982):

$$\frac{\partial}{\partial x}(p + \rho U^2) + \frac{\partial}{\partial z}(\rho \overline{UW}) = 0 \qquad (1.77)$$

where p is pressure, U, W are velocities along the x and z axes, the x axis is directed seaward from the water edge, while the z axis extends vertically

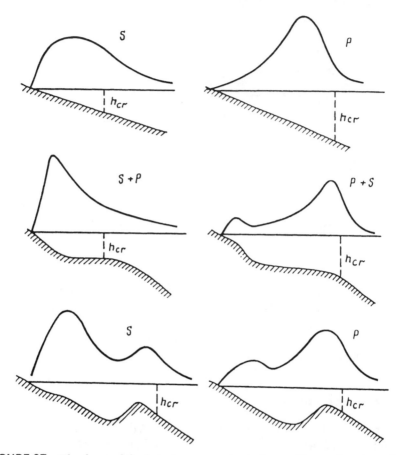

FIGURE 27   The form of the longshore current velocity profile as a function of wave breaking mode and underwater slope profile (Voitsekhovich, 1986a). S, breaking by spilling; P, by plunging; $S + P$, combined mode of wave breaking.

upwards from the calm water level. With allowance for the pressure to be $p = -\rho g(\eta - z) + \rho W^2$, and the velocity presented in the form of $U = \overline{U} + U'$, $W = \overline{W} + W'$ (where $\overline{U}$, $\overline{W}$ are orbital velocities, and U', W' are pulsation components), and assuming turbulence to be isotropic, equation (1.77) can be transformed into

$$\frac{\partial}{\partial x}\rho(\tilde{U}^2 + \tilde{W}^2) + \frac{\partial}{\partial x}\rho g\overline{\eta} + \frac{\partial}{\partial z}\rho\overline{U'W'} = 0 \qquad (1.78)$$

Far from the breaking point, the Reynolds stress defined by the third member of equation (1.78) differs from zero only in a thin bottom boundary layer.

According to Longuett-Higgins (1953), consideration of viscous stresses in the case of a laminar boundary layer gives a three-dimensional pattern of circulation in depth: the upper and lower water layers are moving shoreward, while in the middle layer backrush occurs.

A detailed analysis of the models and experimental data on vertical circulation published before 1980 is given by Leont'ev (1981). We only want to note here that numerous laboratory experiments in smooth horizontal channels or on smooth sloping bottoms without a laminar sublayer, generally confirm theoretical results. Deviations were noted only in the breaking zone of an erodible or non-erodible rough bottom (Bijker, Kalwijk, and Pieters, 1974; Raman, Sethuraman, and Muralikrishna, 1974; Nadaoka and Kondoh, 1982; Hansen and Swendsen, 1984; Stive and Wind, 1986), where strong undertow was always observed below the wave trough. Fig. 28 shows the results of laboratory measurements (Nadaoka and Kondoh, 1982) for various types of wave

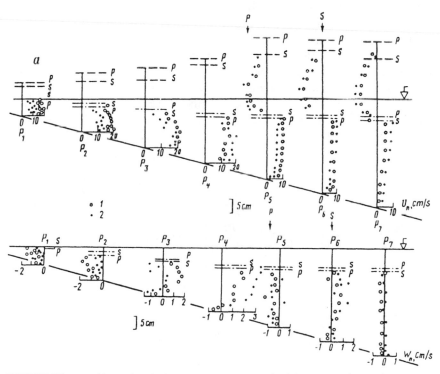

FIGURE 28   Profiles of Euler's mass-transport velocities U_n and W_n for various modes of wave breaking. a) horizontal component; b) vertical component. 1: by spilling (S); 2: by plunging (P); crest position: dashed line; trough position: a dashed dotted line (Nadaoka and Kondoh, 1982).

breaking, illustrating the vertical distribution of mass-transport velocities in various profile points. In areas of wave breaking, a strong undertow is clearly seen. A major water mass is transported shoreward between the wave crest and trough. A part of the velocity profile curve is lacking on the P_1–P_4 vertical lines above the trough, as this area was highly saturated with air bubbles, which made accurate velocity measurements by laser Doppler anemometer impossible. The schematic pattern of circulation based on these measurements is given in Fig. 29. Measurement results obtained by other authors mentioned above support the results shown here.

A qualitative explanation of the vertical circulation pattern for the wave breaking zone is given by Dyhr-Nielsen and Sorensen (1970). According to their concept, at each local level above the bottom, an imbalance between momentum exists due to vertical variations of radiation stress (the first member of equation (1.78)) and due to the constant depth pressure gradient controlled by the flow (the second member of equation (1.78)). This imbalance acts as a driving force of vertical circulation and in a stationary case is held in equilibrium by the Reynolds turbulent stress (the third member of equation (1.78)). This concept was successfully verified by Stive and Wind's experiments (Stive and Wind, 1986) in which the imbalance of the main driving factors of vertical circulation was some 10 times greater within the breaking zone than outside it, where it was practically negligible. This concept formed the basis for the quantitative modeling of surf zone undertow (Dally and Dean, 1984; Swendsen, 1984; Stive and Wind, 1986; Swendsen, Schaffer, and Hansen, 1987). These models are all based on equation (1.78) in which Reynolds turbulent stresses are parameterized by the turbulent viscosity ν_T:

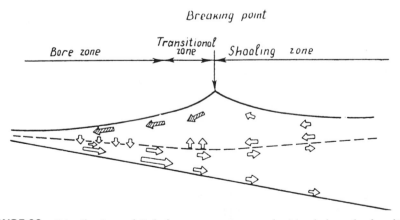

FIGURE 29 Distribution of Euler's mass-transport velocities below the breaking wave (Nadaoka and Kondoh, 1982).

$$\rho \overline{U'W'} = \rho \nu_T \frac{\partial U_n}{\partial z} \qquad (1.79)$$

where U_n is the undertow velocity at elevation z.

The models presented differ mainly in the determination of water discharge transported shoreward above the wave trough and in their approach to the definition of ν_T and selection of boundary conditions. One of the conditions common for all of the models states that to determine the constants for integration of (1.78) with account of (1.79), the mass flow from the bottom to the wave crest section should be equal to zero (zero resulting transport through flow cross-section). Another boundary condition is determined either by a given velocity in the bottom boundary layer, or by a given Reynolds stress at the wave trough level, which depends on the turbulence induced by wave breaking and bore migration due to wave energy dissipation.

In the first numerical model of surf zone undertow, Dally (Dally and Dean, 1984) recognized the mean velocity in the bottom boundary layer as a boundary condition at the bottom and calculated the shoreward mass flow with the help of linear wave theory disregarding the horizontal vortex (see 1.2) contribution to this transport. Swendsen (1984) doubted that linear theory could be legitimately used for transport evaluation in the surf zone, and in his model he allowed for the contribution of bore-organized vortex structures to generate shoreward mass transport, while he assumed the near-bottom velocity to be equal to that at the boundary layer upper limit, in accordance with the Longuett-Higgins model (Longuett-Higgins, 1953). In his later model (Swendsen, Schaffer, and Hansen, 1987), along with the solution of equation (1.78), Swendsen suggested the solution to a problem of the velocity distribution in a turbulent bottom boundary layer for a flat inclined bottom. Detailed discussions of these models in terms of their merits and drawbacks were published in 1986 (Dally and Dean, 1986; Swendsen, 1986).

In contrast to the above models, Stive and Wind (1986) introduced the Reynolds stress at a wave trough level as a second boundary condition. They believed this condition to be more valid physically than a given velocity at the bottom, as the stress is controlled by decrease of the momentum transport above the trough during bore migration. According to the model based on their own experiments, the value of the imbalance (the sum of the first two members in (1.78)) is assumed to be constant in depth below the wave trough.

Comparison of these models with the experimental data described in each of the cited works, gives good agreement. In our opinion, this is mainly achieved by the choice of the constant coefficients present in all the models in varying number.

Swendsen's model (Swendsen, Schaffer, and Hansen, 1987) is generally more physically descriptive as it allows for boundary layer motion and vortex mass transport in a bore, which is not accounted for in other models. Experiments by Japanese researchers (Nadaoka, Hino, and Koyana, 1989) proved this factor to be of great importance. It was shown that the contribution of two-dimensional vortices in the bore area to shoreward mass transport is greater than that of Stokes' transport. Experimental results for a flat sloping bottom enabled them to find a relationship between the wave period averaged mass flow between the wave trough and crest, and the wave period averaged water mass flow by non-vortex orbital motion. Seaward of the wave breaking point, this ratio appeared to be close to unity (as one would expect) while in the surf zone it was 0.75. At the same time, the wave period averaged mass flow was equal to zero in any cross-section of the profile.

The pattern of vertical circulation in natural environments is far more complex, as conditions close to those considered in the models can hardly be created. The intricacy of the bottom relief and swell irregularity nearly always result in three-dimensional nearshore circulation. At the same time, it is extremely difficult to isolate the situation in which the vertical circulation dominates. Thus, only a few attempts have been made to measure vertical circulation in marine environments (Speransky and Leont'ev, 1979; Leont'ev and Speransky, 1979a,b; 1980; Leont'ev et al., 1982; Sallenger et al., 1983; Osborne and Greenwood, 1992a,b). As these measurements of Euler's transport velocities were taken below the trough level, mass flow between the wave trough and crest can not be evaluated in field conditions. Examples of the observed vertical circulation patterns are given in Fig. 30; they generally show qualitative agreement with the above models. Outside the breaking point, water backrush is observed at $kh > 0.5$. In the surf zone the whole water column from the bottom to the wave trough is incorporated in the backrush, just as the above models predicted.

1.5 Infragravity waves

In 1949, Munk (1949) for the first time used the term "surf beat" for sea waves with two- to three-minute periods measured offshore and attributed to height variations of breaking waves. "Surf beat" with 0.5 to 5-minute period is a variety of low-frequency waves in the nearshore zone. A population of all low-frequency waves existing in the nearshore zone in the wind wave range at 0.05 to 0.005 Hz frequency, has recently been called "infragravity waves." Great attention has been concentrated on these waves recently, as they are considered to be one of the main factors controlling the height modulation of onshore wind waves and the spatial scale of the induced circulation which, in its turn, controls the resulting sediment transport and nearshore morphodynamics.

After the breaking of wind waves with 1 to 0.05 Hz frequency, their energy decreases toward the shoreline as water depths decrease, independent of their

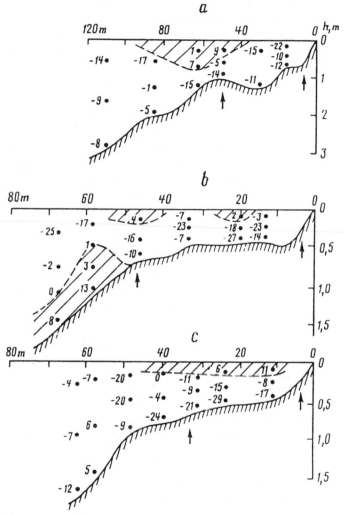

FIGURE 30 Examples of distribution of Euler's mass-transport velocity measured in natural conditions (Leont'ev and Speransky, 1980). The arrow indicates the wave breaking point. The Figs. correspond to values of velocities. The shoreward direction is indicated by +. a) $H_o = 0.57$ m; $\overline{T} = 5.0$ s. b) $H_o = 0.38$ m; $\overline{T} = 4.5$ s. c) $H_o = 0.40$ m; $\overline{T} = 4.5$ s.

pre-breaking energy. Such bore-type breaking waves, in which the height to depth ratio remains constant, are called "saturated" and their energy is completely dissipated near the water edge (Bowen and Huntley, 1984). Infragravity wave energy, on the contrary, grows toward the shoreline, where the greatest morphodynamic modifications usually occur.

Fig. 31 shows the variations of the swash root-mean-square values at frequencies between 0.05 and 0.005 Hz according to root-mean-square ratios of onshore surface waves. It is clearly seen that the amplitude of low-frequency oscillations (<0.05 Hz) grows linearly with the height of onshore wind waves, while the amplitude of the latter remains constant in compliance with the wave saturation concept. Measurements of the swash range on high-energy beaches indicate the predominance of oscillations at <0.05 Hz frequencies (Holman and Bowen, 1984). Modeling these infragravity oscillations is complicated because their frequency and phase structure is unknown and the generation physics is rather complex, being probably related to resonance wave interaction (Battjes, 1988). A short explanation of low-frequency waves is given below, according to Battjes' study.

The whole group of infragravity waves is divided into two main categories: the waves advancing transverse to shore (transversal waves) and edge waves with longshore phase variations. First let us consider the data on transversal waves. Low-frequency waves traveling in a shore-normal direction can be generated by wind waves through various mechanisms. It was shown (Biesel, 1952; Longuett-Higgins and Stewart, 1962) that at constant or slightly varying depths and in approximation to low-amplitude waves, wind waves with a

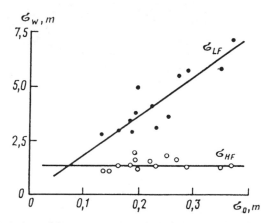

FIGURE 31   Variation of the root-mean-square deviation of swash amplitude (σ_w) at >0.05 Hz (σ_{HF}) frequency and at <0.05 Hz (σ_{LF}) frequency as a function of the root-mean-square deviation of onshore wave level (σ_o) (Guza and Thornton, 1982).

narrow band spectrum can induce a forced long wave, advancing from a group of large waves in such a way that its trough coincides with a group of short-period waves with maximal heights. In compliance with the radiation stress concept, water displacement proceeds below the group of high waves. The amplitude of this long wave is one to two orders lower than that of the short-period wind waves.

Progressive low-frequency waves were registered by Tucker (1950) approximately one kilometer seaward of shore. These waves seemed to correspond to the advance from high-frequency waves observed at that particular place five minutes earlier. This time interval is equal to the time necessary for propagation of the short-period wave group to shore, plus the time of the free long-period wave propagation from shore to the registration point. Long-period seaward waves which are generated, according to Tucker, by high wave groups breaking on the beach, dominated. Thus, the nearshore breaking of a wave group can be a mechanism of low-frequency transversal wave generation. A theoretical model of transversal wave generation above a flat sloping bottom suggested by Symonds and Bowen (1982) was later adopted for a profile with an underwater bar (Symonds and Bowen, 1984). Generation at wave group frequencies is impossible in the surf zone. Seaward of the wave breaking zone, generation is also lacking due to the insignificant dissipation of wave energy. In the exact site of wave breaking, strong dissipation of wave energy proceeds, varying with the wave group, and thus long waves are generated at these group frequencies. Long waves propagate both seaward and shoreward of this local wave breaking zone. That is why they are called "leaky" waves. When migrating shoreward, being reflected, and interacting with onshore waves, they form a system of standing waves transverse to shore. Waves reflected from shore can pass through the generation zone and advance farther seaward. According to this model, the amplitude of low-frequency waves is proportional to the amplitude of the advancing wave group. This model cannot be verified directly by necessary in situ measurements, though its assumptions do not contradict the observed seaward (Munk, 1949; Tucker, 1950) and shoreward (Huntley and Kim, 1984) propagation of transversal waves, and to observed standing transverse waves in the surf zone (Holman, 1981; Guza, Thornton, and Holman, 1984; Sallenger and Holman, 1987).

The second group of nearshore low-frequency waves is formed by "edge" waves. These waves can be progressive or standing. The amplitude of these waves is maximal at the water edge and decreases exponentially seaward. The phase of these waves varies along the shore. In the shore transverse direction they have a modal structure. In approximation to low-amplitude waves above a flat sloping bottom Ursell (1952) showed that the possible edge wave modes are discrete and can be defined by the dispersive relation:

$$\omega^2 = gk[\sin(2n+1)\tan\beta] \qquad (1.80)$$

where ω is angular frequency, $\tan\beta$ is bottom slope, k is wave number, and n is a modal number taking the values of 0, 1, 2… and bounded by $2n + 1 < \pi/2$. In cross section transverse to shore, an edge wave with mode n has n sections of calm water level. The form of edge waves for various modes is shown in Fig. 32.

Generation mechanisms of edge waves are not well-known. Nonlinear resonance interaction of onshore wind waves is considered to be the main generation mechanism of edge waves with periods of less than a few minutes (Gallagher, 1971; Guza and Davies, 1974; Bowen and Guza, 1978; Symonds and Bowen, 1982). Edge waves with periods of up to a few tens of minutes can be formed during storms by atmospheric front transit (Buchwald and de Szoeke, 1973; Worthy, 1984; Middleton, Cahille, and Hsieh, 1987).

Edge wave formation at a subharmonic frequency at reflective beaches with normal and oblique wave approach, predicted by the Guza and Davies (1974) and Bowen and Guza (1978) model was verified experimentally in field conditions (Huntley and Bowen, 1973). Measurements made on a fairly steep beach for low-amplitude waves gave a sharp peak in the velocity spectrum at a typical wave subharmonic frequency. Phase analysis showed that this peak is attributed to the standing edge wave with zero mode.

The role of the surf zone in edge wave generation is still unknown. Abundant in situ data on modes, intensity and structure of low-frequency oscillations in this zone are needed so that new concepts for modeling could arise. Edge waves were intensively studied during the "Nearshore Sediment Transport Study" (NSTS) experiment, with the help of numerous gauges placed in two directions—transverse and along shore at a distance of a few hundred meters (Huntley, Guza, and Thornton, 1981). Longshore velocity measurements showed 1- to 4-minute oscillations produced by progressive edge waves of zero and first modes. The results of shore-normal measurements could not be definitely estimated since long-period waves of unknown origin and source existed along with the edge waves. Edge waves with 40 s to 17-min

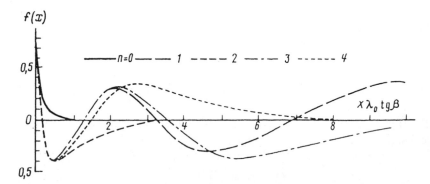

FIGURE 32   Cross-shore profiles of edge waves for various modes (NDCP, 1988).

periods were observed during a storm on the Australian coast near Sydney; the lowest-frequency waves were probably generated by an atmospheric front transit (Middleton, Cahille, and Hsieh, 1987).

As previously noted, infragravity waves play an important role in nearshore relief formation as their length scales are of the same order as many spatial scales of morphological elements observed in natural conditions (Bowen and Huntley, 1984). Transversal low-frequency waves probably play the leading role in the formation of shore-parallel underwater bars. We say "probably" because what we record during in situ measurements is the wave scale on existing profiles with underwater bars and not the generation process of the bar itself with simultaneous measurements of low-frequency wave parameters. It is hard to tell which comes first: either low-frequency waves create a bar at a given distance from shore, or the bar controls the measured length of the low-frequency waves. Examples of such measurements of transversal waves can be found in Sallenger and Holman's work (Sallenger and Holman, 1987).

Bowen and Inman (1971) suggested a model of shore cusps and crescentic bar formation by standing edge waves generated by the interaction of two progressive edge waves similar in mode, frequency, and amplitude but advancing in opposite directions. According to their model, the spatial scales of cusps and crescentic bars alongshore spacing correspond to half the length of the edge wave.

In summary, our knowledge of infragravity wave generation mechanisms needs improvement, especially in environments with intricate bottom relief. Even in a general case of an arbitrary bottom relief, we still cannot determine the main parameters (mode, frequency, amplitude) of infragravity waves. The mutual effect of such waves and bottom relief elements, undoubtedly existing in field conditions, is still unknown. Advance in infragravity wave studies can be achieved thorough laboratory experiments on spatial models where situations with various bottom relief could be simulated or in natural environments for cases approximating model conditions. Access to this information will permit reliable evaluation of existing models while its analysis can possibly result in the generation of new ideas.

CHAPTER

2

Elementary Hydrodynamic Transport Processes

Only a few important phases of hydrodynamic sediment transport in the coastal environment can be easily recognized: the initiation of particle movement, bottom transport, and suspended transport in ripples or smooth flows. The term "elementary processes" was suggested (Longinov and Pykhov, 1981) for the mechanisms characterizing the processes of sediment transport during analysis of nearshore ocean systems. Being the first link and a basis for nearshore studies, elementary processes can be studied both theoretically and experimentally as a physical process of flow, particulate, and bottom interaction.

In natural environments, all the types of sediment transport are generally observed simultaneously, though there are some parts of the underwater slope where one process can dominate. The offshore boundary of sediment transport is at the depth of incipient particle motion for the wave regime. In accordance with sediment coarseness and surface wave parameters, this point can migrate along the slope. When the velocity of near-bottom orbital flow increases, an increasing number of particles become entrapped. In this phase of fluid-sediment interaction, particles are transported as a layer by rolling, sliding, or saltation (with leap height rarely exceeding a few particle diameters), which define bottom sediment transport for a smooth floor. This flow pattern preceding the onset of bottom microform formation is called "the lower smooth phase of sediment motion." Further upslope, velocity growth leads to a greater discharge of transported particles, and under some conditions the formation of ripples begins. These are generally observed on the bottom up to the wave breaking zone. In this phase of flow-bottom interaction, the "ripple" phase, the bottom transport of sand on microform windward slopes proceeds simultaneously with suspended sediment transport in a layer that has a thickness on the order of the ripple length.

Close to the wave breaking point, where waves have a maximum influence on the bottom, ripples are eliminated and sand is dominantly transported as a suspended sheet near the bottom. This flow pattern will be called "the upper smooth phase" of sediment motion. A broad, sheet-like sediment motion above a flat bottom is called "sheet flow." At the exact location of wave breaking, sediment is mostly transported in suspension as a result of vortex flows, while shoreward, from the surf zone with coarser bottom sediment, bedform formation may recommence and the elementary processes can all exist simultaneously. When wave energy is completely dissipated in the swash zone, sheet-like sand motion above a smooth bottom, i.e., the upper smooth phase of sand transport, predominates.

In natural conditions, the intensity of elementary processes and their correspondence to certain zones of the underwater slope depend on the bottom sediment composition, surface wave parameters, and on external forces, such as tidal or setup level variations, wave currents, infragravity waves, and the local bottom slope.

2.1 Initiation of particle movement

One of the main parameters used in quantitative analysis of transported sediment is defined as the condition under which bottom sand particles are disturbed from rest and begin to move in a wave flow. Numerous publications dealing with this problem have appeared lately, but a clear quantitative definition for initiation of particle movement is still lacking. We note two descriptions commonly used for sandy sediments by those studying nearshore and shelf sediment transport: movement initiation proper, when isolated sand grains on a selected bottom area are disturbed from an equilibrium state and begin to move in a water flow, and mass movement of sediment, when the majority of particles on a given surface are moving. A detailed review and analysis of this problem for steady flows and, partially, for wave flow can be found in some monographs (Longinov, 1963; Mirzhulava, 1967; Shulyak, 1971; Rossinsky and Debolsky, 1980; Grishin, 1982; Yalin, 1977; Nielsen, 1979; Fredsøe and Deigaard, 1992) and numerous articles. The main approaches to this problem considered in one of the publications (Beloshapkova et al., 1990) will be discussed briefly below; conditions for initiation of particle movement in a wave flow will be analyzed in greater detail.

Two approaches to the theoretical analysis of critical conditions for particle movement initiation are generally taken. The first one, based on analysis of forces acting on an isolated particle on the bottom, is employed in the majority of publications. The essence of the approach is as follows.

A particle on the bottom is acted upon by the force of gravity $\overline{G}$ and by a force $\overline{F}$ induced by the flow (Fig. 33). The gravity force of a given particle

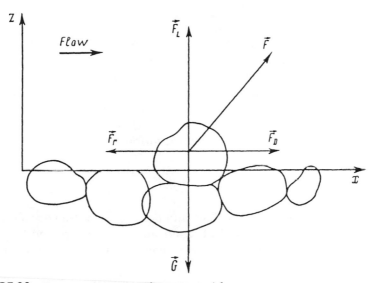

FIGURE 33   Forces acting upon a bottom particle.

remains constant, while the force $\bar{F}$ varies in value and sometimes in direction, depending on flow velocity in the vicinity of a particle and on its geometric properties. This force is usually resolved into two components: lifting force $\bar{F}_L$ and drag force $\bar{F}_D$.

The condition for disturbing a particle at rest and initiation of its movement is determined either from equality of forces acting on a particle, or from equality of their moments. It should be also noted that disturbance of equilibrium of forces or moments should take place in a time interval Δt, necessary for a particle to move far enough from its former position to be able to return to the starting point if the action of sliding or lifting force diminishes.

One of the first who used the approach based on a force balance analysis was Shields (1936). He obtained the criterion for determination of critical conditions of initiation of particle movement in a steady flow by extrapolation of sediment discharge to zero. Critical conditions can be expressed in the form:

$$\Psi_{cr} = \frac{\tau_{cr}}{(\rho_s - \mu)gd} = \frac{\rho U_*^2}{(\rho_s - \rho)gd} = f(\mathrm{Re}_*) \qquad (2.1)$$

where ρ_s and ρ are solid particle and water densities respectively.

Determination of the conditions for initiating particle motion in numerous later studies (Knoroz, 1959; Goncharov, 1962; Mirzhulava, 1967; Mantz,

1977; Yalin and Karahan, 1979) differs only in the methods used for measuring the fluid velocity around the particle and the determination of forces and moments acting upon it. Omitting the details of these studies, we should note that most of them arrive at a form similar to equation (2.1).

In Russian studies, dimensional parameters such as particle diameter and flow velocity are generally used rather than dimensionless parameters for critical conditions. There are a great number of calculation formulas for evaluation of critical conditions. A review of these relations is given by Longinov (1963), Mirzhulava (1967), etc. Most of the formulas are reduced to

$$U_{cr} = A\sqrt{gd} \tag{2.2}$$

where coefficient A is a function of Re_* number, geometric features of the particles, their relative position on the bottom, and other factors, and U_{cr} stands for the depth-averaged flow velocity. Assuming $U_* = aU_{cr}$, equation (2.2) takes a functional form of (2.1).

In the second approach, based on dimensional analysis, particulate mass movement is considered rather than single particle equilibrium. According to dimensional analysis, sediment discharge (q) is a function of four dimensionless parameters defining properties of the particles and fluid (Yalin, 1977):

$$\frac{q}{\rho U_*^2} = \Phi\left[\frac{\rho U_*^2}{(\rho_s - \rho)gd}, \frac{u_* d}{v}, \frac{\rho_s}{\rho}, \frac{h}{d}\right] \tag{2.3}$$

The condition of movement initiation is $q = 0$, i.e.,

$$\Phi_{cr}\left[\frac{\rho U_*^2}{(\rho_s - \rho)gd}, \frac{u_* d}{v}, \frac{h}{d}\right] = 0 \tag{2.4}$$

As there are no inertial forces at the initiation of particle movement, the parameter ρ/ρ_s can be neglected. The parameter h/d is generally disregarded as where $h/d \gg 1$ the flow properties at the bottom are mostly related to near-bottom conditions and do not depend on h. In summary, $\Phi_{cr}(\psi_{cr}, Re_*) = 0$ or $\psi_{cr} = \Phi_{cr}(Re_*)$, and a criterion similar to equation (2.1) is obtained.

Thus, any approach, considering averaged characteristics of flow and sediment, requires determination of the relationship between Ψ_{cr} and Re_*, which can be based only on experimental data. Shields was one of the first who did it for uniform particles. The results of his experiments are shown in an insertion to Fig. 34. Results of later studies are shown in Fig. 34 and it follows from them that:

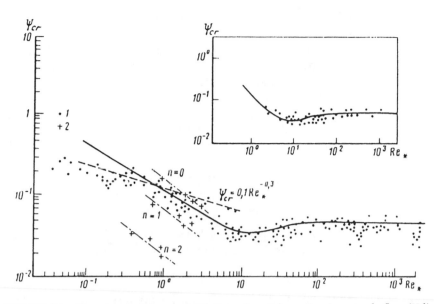

FIGURE 34   Experimental data on initiation of particle motion in a steady flow (Yalin and Karahan, 1979). Shields' experimental data are shown in an insertion.

1. For a completely turbulent regime $Re_* \geq 70$, $\Psi_{cr} = const = 0.05$
2. For a hydraulically-smooth regime $Re_* \leq 8$ experimental data are approximated fairly well by rather simple relationships, e.g., by the Muntz formula

$$\Psi_{cr} = 0.1 Re_*^{0.3} \qquad (2.5)$$

3. According to most authors, the minimum of Ψ_{cr} is observed in the transitional area.

Though abundant experimental results are available, their correct comparison in the form of equations (2.1) or (2.2) is hardly possible due to the lack of necessary information or different approaches for determining critical conditions. In general, the scatter of Ψ_{cr} values in relation to an approximating curve is around 50%. However, as each point represents the result averaged from a few measurements (generally from 5 to 20) the real scatter is on the order of the measured value.

Inaccurate determinations of critical conditions result not only from subjectivity, but from objective features of the flow-particle interaction which cannot be allowed for when critical conditions of initiation of sediment movement are based on averaged parameters. These features can be allowed for if the process is analyzed by instantaneous characteristics, i.e., with the help of probabilistic-statistical methods. Let us next consider the main aspects of this approach.

For a surface layer of sediment, the local shear stress necessary to displace a particle at rest will depend on particle properties. Thus, there is some probability distribution of critical tangential stresses $P(\tau_{cr})$ or of critical velocities $P(U_{cr})$ for any type of sediment. Tangential stresses and velocities in near-bottom fluid flow have their own distributions $P(\tau)$ and $P(U)$. If the conditions of $\tau > \tau_{cr}$ and $U > U_{cr}$ are fulfilled for a particle group, then critical conditions for sediment movement are defined by the degree of intersection of $P(\tau)$ with $P(\tau_{cr})$ or $P(U)$ with $P(U_{cr})$ (Fig. 35). The degree of this intersection can be found from the probability of sediment particle motion. Rossinsky (1968) used this method for uniform particles. Later (Beloshapkova et al., 1990) it was extended to nonuniform sediment with the following assumptions:

1. The sediment grain size distribution is lognormal.
2. Deviation of the instantaneous near bottom velocity U from the averaged velocity $\overline{U}$ obeys the normal law. $\sigma_U/\overline{U}$ ratio varies insignificantly from 0.2 to 0.3.
3. The equation for critical velocity U_{cr} is in the form of equation (2.2).

For uniform bottom sand, the dependence between take-off probability and flow velocity can be written in a general form as

$$\eta = \frac{1}{\sqrt{2\pi}\sigma_u} \int_{-\infty}^{\infty} \exp-\left[\frac{(U-\overline{U})^2}{2\sigma_U^2\overline{U}}\right] P(U_{cr})dU \qquad (2.6)$$

where $P(U_{cr})$ is a function of the distribution of take-off critical velocity, for uniform bottom

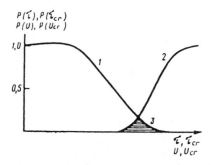

FIGURE 35 Determination of critical conditions for initiation of sediment motion: 1) function of $P(\tau)$ and $P(U)$ distribution; 2) function of $P(\tau_{cr})$ and $P(U_{cr})$ distribution; 3) an area in which conditions for particle motion are satisfied.

$$P(U_{cr}) = 0 \text{ at } U < U_{cr} \tag{2.7}$$

$$P(U_{cr}) = 1 \text{ at } U \geq U_{cr}$$

For nonuniform bottom $P(U_{cr})$ is defined by the relationship $U_{cr} = f(d)$ if assumptions 1-3 are fulfilled:

$$P(U_{cr}) = \frac{2}{\sqrt{2\pi}\sigma_{\ln d}} \int_0^U \exp\left\{-\frac{[2\ln(a/m)]^2}{2[\sigma_{\ln d}]^2}\right\} \frac{1}{a} da \tag{2.8}$$

where $m = a\sqrt{gd_{50}}$ and d_{50} is the median particle size. Calculation results are shown in Fig. 36.

Comparison of modeling results with laboratory experiments (Dobroklonsky, Mikhailova, and Mylukova, 1976) and in situ observations (Shtelzer, 1984) shows that the model gives slower growth of η against U, which is caused mainly by limitations of the model itself. Adoption of a log-normal distribution of sediment grain size actually corresponds to particle take-off from a flat surface, where sediment grains are situated at some distance and escape mutual influence. In real conditions, however, effects of particle shielding, seizure, and self-paving are unavoidable, so that the relationship can be far from lognormal.

Grass (1970) was one of the first who measured instantaneous values of bottom tangential stresses τ_{cr} corresponding to various degrees of sediment movement, from the first displacement of singular grains to mass sediment transport and bedform formation. In his work, the degree of intersection of $P(\tau)$ and $P(\tau_{cr})$ is defined by the parameter n in a condition (Fig. 37)

$$\bar{\tau} + n\sigma_\tau = \bar{\tau}_{cr} - \sigma_{\tau_{cr}} \tag{2.9}$$

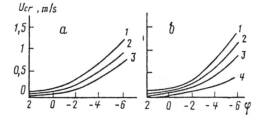

FIGURE 36   Dependence of threshold velocity of particle motion on particle diameter and take-off probability. (a) 1: $h = 5\%$; 2: -1%; 3: -0.1%; $A = 3$, $s_j = 1$, $s_u/\bar{U} = 0.25$ and composition heterogeneity. (b) 1: $s_j = 0.2$; 2: -0.5; 3: -1; 4: -2; $A = 3$; $h = 1\%$, $s_u/\bar{U} = 0.25$.

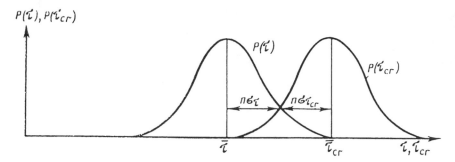

FIGURE 37   The critical conditions for initiation of particle motion (Grass, 1970).

Experimenting with fine sand grains with mean diameter from 0.075 mm to 0.21 mm, Grass obtained $\sigma_t = 0.4\bar{\tau}$ and $\sigma_{\tau cr} = 0.3\bar{\tau}_{cr}$. Thus equation (2.9) can be written in the form:

$$\bar{\tau} = \frac{1-0.3n}{1+0.4n}\bar{\tau}_{cr} \tag{2.10}$$

The experiments showed that at $n = 2$, bottom particles were all stable, at $2 > n > 0$ particles moved with varying degrees of intensity, and at $n = 0$ bed forms were formed. These data plotted against Shields' diagram (Fig. 34) indicate that point scatter is due to the authors' subjectivity in defining the initiation of particle movement (from the first displacement of particles to formation of bed forms).

It can also be shown that in a transitional and hydrodynamically rough regime, variations of Ψ_{cr} are of a similar range (Fig. 38). Thus, in Fenton and Abbott's experiments (Fenton and Abbott, 1977) with a variable value of D/d (where D is the mean height of particle projection above the bottom) Ψ_{cr} varies by an order of magnitude, while the Ψ_{cr} value obtained by Shields' curve corresponds to $D/d = 0.1$, i.e., it is nearly the same as in Shields' experiments. Coleman's experimental data (Coleman, 1967) obtained for large D/d values, show a similar discrepancy in Ψ_{cr} estimates, which is also confirmed by results of in situ observations (Hammond, Heathershaw, and Longhorne, 1984) carried out on a natural gravel-pebble bottom. This implies that for completely turbulent flow in natural conditions, Ψ_{cr} values vary from 0.5 Ψ_{cr} to 2 Ψ_{cr}, so that possible errors in Ψ_{cr} values taken from Shields' curve can be of the order of the measured value.

Up until now, our discussion was confined to steady flow. Similar studies for wave flow are scarce, but general aspects of the problem have much in common. It is frequently concluded that the critical velocities for wave and steady flows are similar. At the same time, it is not elucidated which velocities

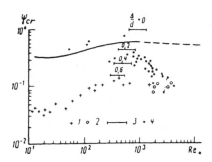

FIGURE 38   Comparison of Coleman's (1967), Fenton and Abbott's (1977), Hammond's (Hammond et al., 1984) data with Shields' curve. 1: Coleman, 1967; 2: Fenton, Abbot, 1977, for $D/d = 0.82$; $D/d = 0.6$; 3: for $D/d = 0.6$; 4: Hammond et al., 1984.

are considered for wave (peak or average) and steady (depth-average or near-bottom) flows. We find this comparison incorrect, as it holds true only in the case of instantaneous values of τ_{cr} and U_{cr}. Empirical formulas for determining particle movement initiation on a smooth bottom were suggested in the first fundamental laboratory studies (Bagnold, 1946):

$$\frac{\omega^{1/4} U_m^{3/4}}{\left(\frac{\rho_s - \rho}{\rho}\right)^{1/2} \bar{d}^{0.325}} = 21.5 \tag{2.11}$$

and (Manohar, 1955):

$$\frac{U_m}{\left(\frac{\rho_s - \rho}{\rho}\right)^{0.4} (v\bar{d})^{0.2}} = 8.2 \tag{2.12}$$

In Bagnold's formula, parameter values are taken in the CGS system. These formulas are rarely used now, as the parameter range used for their derivation is limited.

Komar and Miller (1973) suggested a criterion widely used now for evaluation of movement initiation. For sand grains smaller than 0.05 cm they suggested

$$F_m = \rho U_m^2 / (\rho_s - \rho) g d = a'(a_m / d)^{0.5} \tag{2.13}$$

where a' is a dimensionless coefficient found empirically, and F_m is a mobility factor. From experimental data of Bagnold (1946) and Manohar (1955) they found $a' = 0.30$. Critical analysis of Bagnold's and Manohar's data and application of results obtained by other authors caused Komar and Miller (1975) to reduce the a' value to 0.11.

Rigler and Collins (1983, 1984) used the relationship (2.13) for comparison with results of their own studies of initial conditions of movement of mineral particles with various specific weights: quartz sand, ilmenite, and cassiterite. They found out that for particles with diameters 0.14 to 0.28 mm, equation (2.13) describes the observed conditions of movement initiation with 10% accuracy if a constant value of a' is assumed to be 0.32.

Formula (2.13) served as a basis for in situ studies of initial conditions (Sternberg and Larsen, 1975). To calculate wave parameters, waves of 10% maintenance were considered. Comparison of observed initiation of particle movement with calculations by (2.13) gave $a' = 0.13$, which practically corresponds to the value reported by Komar and Miller (1975).

Madsen and Grant (1976) suggested using Shields' criterion for characterization of initial conditions for sand movement in a wave flow. They showed that if τ_{cr} in (2.1) is found from (1.9), Shields' curve for steady flow can be used for evaluation of initial conditions under a wave. For convenience of practical calculations, it was modified by plotting

$$\mathrm{Re}_{\overline{d}} = \frac{\overline{d}}{4\nu\sqrt{(\rho_s/\rho - 1)g\overline{d}}}$$

on the abscissa (Fig. 39). Vertical lines in this figure show the range of initial conditions defined in all known experimental studies published before 1976. In natural nearshore environments, quartz sand usually predominates with

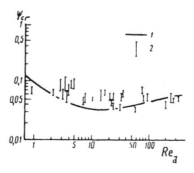

FIGURE 39   Modified Shields' curve (1) and experimental data on initiation of particle motion in a wave flow (2) (Madsen and Grant, 1976).

grain diameters from 0.1 to 1 mm. For such sands, Re varies from 1 to 90, and the critical value of Shields' parameter varies from 0.03 to 0.1.

In later works of Dingler (1979) and Hallermeier (1980), empirical criteria of movement initiation based on observations and on analysis of published results were suggested. However, as these formulas include a number of empirical constants, they are not used widely. To define the initial conditions of wave-induced particle movement on a shelf or in the nearshore zone, most current researchers use either (2.13) or a modified Shields' curve (Fig. 39).

Let us consider those few in situ observations that concern the initial conditions under irregular waves. Larsen with co-workers (Larsen et al., 1981) made simultaneous observations of silty sand bottom sediment and detailed instrumental measurements of near-bottom velocities on the US and Australian shelves. One meter above the bottom, current velocities were observed to not exceed 20 cm/s and were less than the orbital velocities. In these conditions, a modified Shields' plot appeared to give good predictions of initial conditions of particle movement. Estimates by (2.13) fall within the limits of scatter of the experimental data on which this equation is based.

Thorough and detailed observations of the initial conditions of wave-induced sand movement were made by British scientists (Davies and Wilkinson, 1978; Davies, 1980). Observations of initial sand displacements were taken with a special TV camera placed close to the bottom and were thoroughly synchronized with measurements of the horizontal components of orbital velocity at 1 m above the bottom. The observed bottom area was covered by bed forms symmetric about the crest, 12 cm high and 85 cm long. These forms, created by a previous wave regime, were passive. Locally, these forms were rubbed off by divers so that the initial conditions on a flat bottom could be studied. The bottom sediment was mainly composed of 1- to 2-mm diameter particles, the mean diameter being 1.4 mm. In the course of observation of a rippled bottom, sand particle movement was confined to the ripple crests, behind which no vortices were formed. The actual initiation of sand movement and its direction were registered for each wave half-period.

Histograms of orbital velocity amplitudes are shown in Fig. 40, where dark color is used only for situations with initial displacement of bottom particles. In both cases, the swell period was about 9 seconds. In the case of a rippled bottom (Fig. 40a) sand movement begins only at $|U_m| > 8$ cm/s, being induced only by some of waves at $8 < |U_m| < 22$ cm/s, and by all of them at $|U_m| > 22$ cm/s. In the case of the smooth bottom and under the same conditions, initial movement of particles is observed only at $|U_m| \gg 30$ cm/s. Though the amplitude of the orbital velocity at 1 m above the bottom rarely exceeded 30 cm/s during the observation period, artificial erasure of bed forms results in the complete damping of bottom sediment movement in contrast to the rippled bottom. Critical values of orbital velocity amplitudes are nearly two times lower for a rippled bottom than for a smooth one. This

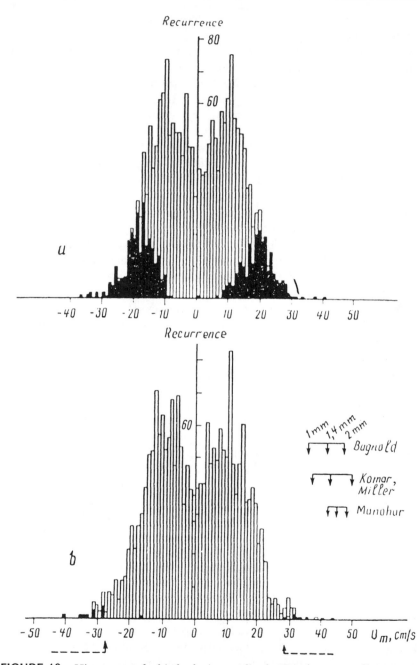

FIGURE 40 Histograms of orbital velocity amplitudes. Displacement of bottom particles shown by black color. Positive values of velocity: shoreward; negative: offshore. a) rippled bottom; b) flat bottom (Davies and Wilkinson, 1976).

fact has been repeatedly noted in laboratory conditions (see, e.g., Shulyak, 1971). It is explained by the fact that the near-crest velocity due to current bending, or convective acceleration, is nearly twice as high as that on a smooth bottom, water depth, and surface wave parameters being equal (Du Toit and Sleath, 1981). Critical values of velocity amplitudes estimated by the criteria of Bagnold (1946), Manohar (1955), and Komar and Miller (1975) for particle diameters of 1, 1.4, and 2.0 mm are shown in Fig. 40b. These estimates generally agree with measurements. Davies showed in his study (Davies, 1980) that good agreement with this data is achieved by calculations based on the curve in Fig. 39.

It should be noted that apart from the above factors (sediment nonuniformity, existence of passive bed forms or other types of roughness), the stabilizing and destabilizing effect of benthic organisms also plays an important role in estimates of the initial conditions of bed sediment take-off (Grant, Boyer, and Sanford, 1982; Nowell, Jumars, and Eckman, 1981). Secretions from benthic organisms change the physical-chemical properties of surface sediment, while mounds and holes built by them cause local variations of the velocity field. The problem of determining initial conditions in these situations and the possible use of the above criteria for their estimates are discussed in the studies cited. Unambiguous recommendations are still lacking, and many factors (such as benthic organisms species, habitat composition, season, etc.) seem to be important. The order of magnitude can still be found from the Shields' curve.

Great difficulties arise when estimating the initial conditions of shell movement as the typical diameter included in the above criteria cannot be derived from their shape. Instead, a sphere with an equivalent diameter, density, and settling velocity must be used to make estimates. Experimental studies of detrital material showed that if this approach is taken, initial conditions can be estimated using the Shields' curve for a first approximation (Fisher, Pickral, and Odum, 1984).

2.2 Bed microforms

When the near-bottom velocities of water flow slightly exceed their values for the mass transport of bottom sediment, periodic forms of microrelief are formed. Depending on the intensity of the near-bottom flow, sediment composition, and nature of the bed surface, various types of bed microforms may be built. Fig. 41 shows their classification after Sleath (1984). The classification includes the main types of bed forms observed in laboratory and field conditions for a purely wave flow. Breaking zone bed forms ignored by this classification will be discussed in Chapter 5. Observations of these forms are rare and generally qualitative. To place them in this classification, detailed

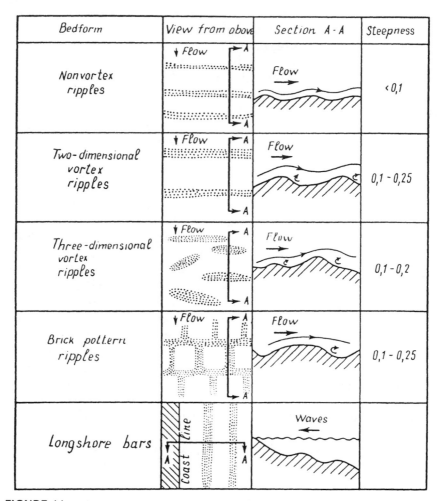

Bedform	View from above	Section A-A	Steepness
Nonvortex ripples	Flow	Flow	<0,1
Two-dimensional vortex ripples	Flow	Flow	0,1 - 0,25
Three-dimensional vortex ripples	Flow	Flow	0,1 - 0,2
Brick pattern ripples	Flow	Flow	0,1 - 0,25
Longshore bars	Coast line	Waves	

FIGURE 41   Classification of wave flow bedforms (Sleath, 1984).

data on bedform parameters and the physics of their formation are necessary. The classification in Fig. 41 also includes underwater bars, the geometric dimensions of which are by a few orders of magnitude greater than that of other bed forms. It seems that underwater bars cannot be classified as purely wave-induced relief forms, as their construction is not directly related with the purely orbital near-bottom flow induced by surface waves but most probably is caused by the infragravity waves discussed in Chapter 1. Therefore, the analysis of bed microforms will be confined to those formed by the interaction between surface waves and an erodible bottom.

2.2.1 Rolling grain ripples

Bagnold was the first who observed ripples of this type (Bagnold, 1946) at flow velocities insignificantly higher than critical values of particle movement initiation. Separate grains roll off from stable positions and, due to oscillatory water movements, build up into a series of parallel rows normal to the flow. According to Carstens and Neilson (1967), the bottom surface at this stage is separated into parallel bands with rolling sand grains and areas with stationary particles. Particle rolling is confined to these bands and is in phase with the near-bottom orbital velocity. The crests of rolling grain ripples are formed by moving sand bands, while their troughs correspond to stationary areas. As grain rolling is the dominating type of sediment transport at this stage of wave-bottom interaction, Bagnold (1946) called the constructed bed forms "rolling grain ripples," i.e., ripples formed by rolling grains.

The profile geometry of rolling grain ripples depends on the diameter of the bottom sediment: it approximates a sine curve for coarse particles, while for fine particles the troughs are always flat. Laboratory studies (Bagnold, 1946; Manohar, 1955; Carstens and Neilson, 1967; Sleath, 1976) of a mobile bottom showed these ripples to be stable over a wide range of near-bottom orbital velocity amplitudes: from that of first grain displacement to twice as high. Experiments in wave tunnels showed their instability and rapid transformation into vortex ripples; this fact still requires a satisfactory explanation.

Some of the authors consider the low steepness, $H_r/l_r < 0.1$, of rolling grain ripples to be a main characteristic (Miller and Komar, 1980a; Sleath, 1984), as at low steepness there is no separation of the boundary layer around the crest and a vortex roller is not formed. This limitation of steepness seems to be meaningful only for two-dimensional forms, i.e., when their crests are constant in height and parallel to each other, being normal to the wave direction.

Low steepness of sand microforms is also observed at the point of vortex ripple disappearance. In this situation vortices are formed around the crest even with low steepness due to large near-bottom velocities. In this phase of wave-bottom interaction, ripples generally become three-dimensional (their height varies along crest), and large volumes of sand are transported around them in suspension. Therefore, we classify two-dimensional bed forms with steepness lower than 0.1 as rolling grain ripples, the existence of which is not accompanied by sediment suspension. We can hardly agree with Sleath (1984) who classified microforms with steepness less than 0.1 existing in the phase of ripple disappearance and formation of the upper smooth phase of sediment movement as rolling grain ripples.

The physical pattern of rolling grain ripple formation is still vague. Thorough visual observations of sand and glass balls transport (Carstens and Neilson, 1967; Kaneko and Honji, 1979) showed that particle rolling is confined to the boundary layer, though the physics of a collection of spontaneously scattered particles into parallel bands is still unknown. Kaneko and Honji (1979) somewhat inconclusively explain it by interaction forces between two close particles passed over by a viscous liquid. According to Sleath (1976, 1984), development of these forms can be explained by peculiarities of mass transport in the vicinity of an undulated bottom surface or close to an artificial ridge on a flat bottom. He showed that particle displacement and the development of rolling grain ripples are controlled by the constant water flow from trough to crest.

Sleath (1976) and Kaneko (1981) tried to predict rolling grain ripple length theoretically, but these predictions are purely estimative as the suggested formulas include a number of empirical constants.

2.2.2 Vortex ripples

With greater flow velocity above a bottom covered by rolling grain ripples, or with a high ridge or a deep trough existing around them, bottom boundary layer separation occurs, and vortices leeward of the crest or ridge are formed. Eddies with a horizontal axis cause local bottom erosion resulting in ripple height growth as sediment is carried toward its crest. If the wave regime remains unchanged for some time, the characteristics of two-dimensional ripples become stable and are well preserved for a long time. Laboratory observations show (Sleath, 1984) that with coarser sand, the crests of stable vortex ripples become rounded and approximate a sine curve. If the bottom is covered by fine sand, the crests of vortex ripples have a sharp form.

Under purely periodic water flow, vortex ripples are symmetric about the crest. Two-dimensional ripples have maximum steepness of 0.15–0.25 under an equilibrium regime. The pattern of sand movement in the vicinity of two-dimensional ripples is shown in Fig. 42 which illustrates suspension cloud formation and motion in various phases of near-bottom orbital flow. Particle displacement by rolling and saltation is observed on windward ripple slopes.

As the flow velocity grows, two-dimensional ripples gradually become three-dimensional and their steepness decreases. The conditions under which this transformation is observed are discussed below.

As mentioned in the beginning of this section, vortex ripples are formed due to velocity growth either from rolling grain ripples formed earlier on a flat bottom, or when bottom ridges or troughs are high enough to create vortices in their vicinity. In the latter case, vortex ripple formation occurs at a lower amplitude of near-bottom orbital velocity than that needed for initiation of particle movement on a flat bottom. When an artificial bed ridge exists, the bottom is eroded in the vicinity of the vortex roller with sediment transported toward it in each wave half-period. Gradually, ripples develop on both sides of the ridge. When the first ripple crest has been formed, a vortex roller develops behind it and the process is repeated. A detailed discussion of

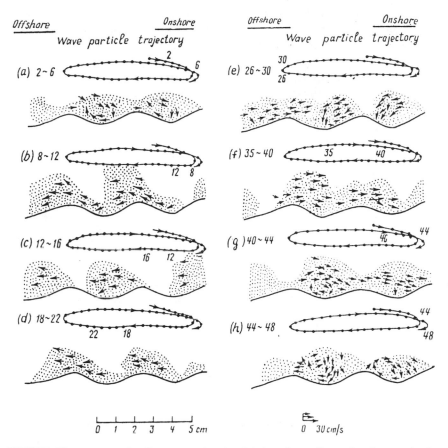

FIGURE 42   Pattern of sediment motion in vicinity of two-dimensional wave ripples (Sunamura, Bando, and Horikawa, 1978; NDCP, 1988).

the vortex ripple formation mechanism can be found in a number of studies (Carstens and Neilson, 1967; Sleath, 1984; Shulyak, 1971).

Theoretical studies of vortex ripple formation are practically absent since a precise definition of sediment movement is too complicated in this case. The only exception is the work by Sleath (1975), in which vortex ripple lengths are estimated by a formula based on the assumed laminar flow, the low intensity of sediment transport, and the absence of sediment exchange between two neighboring ripples. The calculations are in good agreement with laboratory data. Data on the geometrical parameters of vortex ripples and their variations are mainly obtained in laboratory experiments (Bagnold, 1946; Manohar, 1955; Yalin and Russel, 1962; Kennedy and Falcon, 1965; Carstens, Neilson, and Altinbilek, 1969; Mogridge and Kamphuis, 1972; Lofquist, 1978;

Sleath and Ellis, 1978; Nielsen, 1979; Kaneko and Honji, 1979; Shulyak, 1971) and in situ observations (Inman, 1957; Dingler, 1974; Dingler and Inman, 1976; Miller and Komar, 1980a; Kos'yan, 1985; 1988a). Some of these data are generalized by Nielsen (1979, 1981). Some of his results are discussed below.

To define the geometric characteristics of vortex ripples, Shields' parameter for wave flow and the mobility parameter F_m are usually employed:

$$F_m = \frac{\rho U_m^2}{(\rho_s - \rho)g\overline{d}} = \frac{\rho(a_m\omega)^2}{(\rho_s - \rho)g\overline{d}} \tag{2.14}$$

which are interrelated as

$$\Psi = \frac{1}{2}f_w F_m \tag{2.15}$$

When the bottom is rough, which is the case in natural conditions, vortex ripples are formed at $\Psi = 0.045$. Their steepness depends on the Shields' parameter and on the sediment friction angle Φ. At $\Psi < 0.2$ ripple steepness is maximal, independent of Ψ, and varies in an interval:

$$0.25\Phi < \frac{H_r}{\lambda_r} < 0.5\Phi \tag{2.16}$$

At $0.2 < \Psi < 1$ ripple steepness decreases and at $\Psi \gg 1$ they have disappeared. The mobility factor is often used for defining the ripple length. At $F_m < 20$ the ripple length is of the same magnitude as the near-bottom fluid oscillations:

$$\lambda_r / a_m = const \tag{2.17}$$

The value of this constant varies in different experiments from 1.0 to 2.0. Ripple geometry can be estimated by empirical formulas derived by generalization of abundant laboratory and in situ measurements, suggested by Nielsen (1979, 1981). For laboratory conditions:

$$\frac{\lambda_r}{a_m} = 2.2 - 0.345 F_m^{-0.34} \quad \text{for} \quad 2 < F_m < 230 \tag{2.18}$$

$$\frac{H_r}{\lambda_r} = 0.182 - 0.24\Psi^{1.5} \tag{2.19}$$

$$\frac{H_r}{a_m} = 0.275 - 0.022 F_m^{0.5} \quad \text{for} \quad F_m < 156 \tag{2.20}$$

For field conditions:

$$\frac{\lambda_r}{a_m} = \exp\left[\frac{693 - 0.37\ln^8 F_m}{1000 + 0.75\ln^7 F_m}\right] \tag{2.21}$$

$$\frac{H_r}{\lambda_r} = 0.348 - 0.34 \sqrt[4]{\Psi} \tag{2.22}$$

$$\frac{H_r}{a_m} = 21 F_m^{-1.85} \tag{2.23}$$

Nielsen also notes good agreement between in situ and laboratory measurements, if a_m and U_m are calculated by significant wave heights. From the Russian studies, the formula derived by laboratory experiments (Altunin, 1975) should be noted:

$$H_r = 2.48 \cdot 10^{-3} \frac{U_m}{\sqrt{gd}} \lambda_\gamma \tag{2.24}$$

The mechanism of transformation of two-dimensional ripples into three-dimensional ones is very poorly studied. Carstens with co-workers (Carstens, Neilson, and Altinbilek, 1969), on the basis of wave tunnel experiments, found this transformation to occur at $a_m/\bar{d} \leq 775$. This corresponds to the boundary between the zone with $\lambda_r/a_m = const$ and the zone where λ_r/a_m decreases with F_m growth. In these experiments a constant period of 3.56 s was used.

Now let us consider the limits of ripple existence. It is generally accepted that ripples are formed at velocities only slightly higher than that at movement initiation. Manohar (1955) estimated this velocity ratio as 1.2; by other estimates it is 1.16 (Carstens, Neilson, and Altinbilek, 1969). According to Nielsen (1979), ripples develop at $\Psi = 0.045$, while Komar and Miller use the equation (2.13) with $a' = 0.11$ for determination of initial conditions.

In some works, the initial conditions for ripple formation are defined with the help of the mobility parameter,

$$F_m = 3 \text{ (Brebner, 1980)} \tag{2.25}$$

$$F_m = 0.89\left(\frac{2a_m}{\bar{d}}\right)^{1/3} \quad \text{(Vongvisessomjai, 1984)} \tag{2.26}$$

$$F_m = 0.04\left(\frac{2a_m}{\bar{d}}\right)^{-2/3}\left[\frac{gd^3(\rho_s - \rho)}{v^2\rho}\right]^{1/9} \quad \text{(Dingler, 1979)} \tag{2.27}$$

Estimation of these formulas using in situ observations is given in Chapter 5.

The upper limit of ripple existence is defined by the onset of the upper smooth phase of movement, during which bedforms disappear and suspended sediment transport becomes very intensive. According to laboratory experiments (Carstens and Neilson, 1967), ripples disappear at

$$a_m/\bar{d} = 17,000 \tag{2.28}$$

while Kennedy and Falcon (1965) on the basis of the analysis of Inman in situ measurement data (Inman, 1957) got the value:

$$a_m/\bar{d} = 8,000 \tag{2.29}$$

By in situ measurement results Dingler and Inman (1976) defined the upper limit of ripple existence by the mobility factor as:

$$F_m = 240 \tag{2.30}$$

Analyzing the effect of wave period, Vongvisessomjai (1984) obtained:

$$F_m = 12.7\left(\frac{2a_m}{\bar{d}}\right)^{1/3} \tag{2.31}$$

The Shields' parameter for wave flow is most frequently used for determination of the upper limit of ripple existence:

$$\Psi = 1 \quad \text{(Nielsen, 1979)} \tag{2.32}$$

$$\Psi = 4.4\left(\frac{U_m\bar{d}}{v}\right)^{-1/3} \quad \text{(Komar and Miller, 1975)} \tag{2.33}$$

$$\Psi = 0.5 \text{ to } 0.6 \quad \text{(Horikawa, Watanabe, and Katori, 1982)} \tag{2.34}$$

The interval of ripple existence in Ψ and $\overline{d}/4\nu\sqrt{[(\rho_s/\rho) - 1]g\overline{d}}$ coordinates is shown in Fig. 43. The validity of the suggested criteria is analyzed in Chapter 5 in the discussion of in situ observation data.

2.2.3 "Brick pattern"-type ripples

Bagnold (1946) was the first who observed these ripples in his experiments. They were formed at low near-bottom water oscillations, $a_m < L_0/6$, where L_0 is the maximum length of vortex ripples which could be formed with the given composition of bottom sediment. In the formation mechanism, they are similar to two-dimensional vortex ripples with the difference that in this case, two-dimensional eddies disintegrate into a series of horseshoe shaped eddies (Fig. 44) which form lengthwise crests between two-dimensional ripple crests. Sleath (1984) notes two factors that limit development of these forms in laboratory conditions. First, the distance between lengthwise crests should be of the same order as the distance between two-dimensional ripple crests, so in narrow flumes, brick-pattern-type ripples are not observed. The second limitation is related to sediment discharge. These forms begin to disappear under the intensive transport of suspended sediment, when particles suspended from the crests are transported by vortices and lose constant contact with the bottom. In Sleath and Ellis' experiments (Sleath and Ellis, 1978) such forms were developed at $a_m < L_0/2$, and the authors note that at $a_m > L_0/2$ they did not disappear abruptly, but could be followed on the bottom for some time.

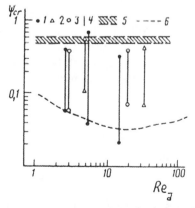

FIGURE 43 Area of wave ripple existence (Horikawa, 1988). Experimental data: 1: Carstens, Neilson, and Altinbilek, 1969; 2: Manohar, 1955; 3: Shibayma, 1984; 4: zone of form existence; 5: Horikawa, Watanabe, and Katori, 1982, boundary of the upper smooth phase of sediment motion; 6: modified Shields' diagram for wave flow (Madsen and Grant, 1976).

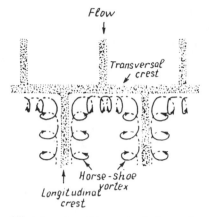

FIGURE 44   Scheme of "brick pattern"-type ripple formation (Sleath, 1984).

2.3 Suspended sediment

We have noted in the above paragraphs that with the growth of near-bottom velocity, a greater amount of suspended sediment is transported by flow. During the development of rolling grain ripples, the near-bottom transport of sand particles predominates. As velocity increases, sediment transport goes through the phases of growth of vortex ripple development, transformation into three-dimensional forms, and the disappearance and transition to the upper smooth phase of movement that finally becomes the dominating type of sediment transport. In field conditions, this is observed close to the wave breaking point, where longshore currents have the highest velocities. Thus, prediction of the suspended sediment concentration profile in various phases of bottom development becomes one of the key aspects of the sediment transport problem. Before we turn to transportation of suspended sediment by flows of specific types, we will name the principal theoretical approaches to the description of suspended flows.

2.3.1 Theoretical models of suspended flows

For convenience of presentation, all models are considered in groups according to the physical premises used for their construction.

2.3.1.1 Movement of a single solid particle. It seems logical that the theoretical study of the motion of particulate sediment suspended in water flow should consist of the study of the motion of a single particle with further extension to the motion of a group of particles. The solution of the motion of a single particle provides valuable information that can be used for clarification of suspension mechanisms and for determination of sediment discharge caused by the presence of solid particles in liquids.

Particle behavior in a turbulent flow depends largely on the size of this particle in comparison with the scale of turbulence. If a particle is rather large, turbulence will mainly cause its greater resistance to flow and the particle will, at most, follow the large-scale turbulent motion of the liquid. But if a particle is small in comparison with the lowest scale of turbulence, it will tend to follow all the components of liquid turbulence.

Most theoretical schemes considering the behavior of a single particle placed in a given liquid volume assume it to be small enough to execute motions fully imitating the turbulent pulsations of this volume. Under this limitation (i.e., with reference to fine particles), theoretical studies of suspended particle motion were carried out and discussed by Granat (1960), Shraiber (1973a), Kachanov and Levich (1967), Dzrbashyan (1963), Tchen (1947), Soo (1956), Liu Vi-cheng (1956), Friedlander (1957), Chao (1964), Hjelmfelt and Mocros (1966), Bailey (1974), Smutek (1974), etc.

Having reviewed these studies (a detailed analysis is given by Kos'yan, 1983a,b; Antsyfcrov and Kos'yan, 1986) one has to admit that this approach is far from practical. A detailed description of solid particle motion in viscous fluid, allowing simultaneously for the effect of the liquid velocity field on body motion and for the inverse effect, at present is still lacking. The studies under review generally consider the motion of very fine particles. It is generally assumed that particle resistance to motion in liquid depends linearly on their relative velocity. Calculation methods allowing for the square resistance law which are more significant for practical use (Haskind, 1956; Panchev, 1959), are too complex, and even in the first approximation, result in huge formulae inconvenient to use. All the studies assume a particle to be staying within a flowing volume of liquid for a rather long period. This assumption, even for short time intervals, is generally unsatisfied (Hinze, 1963; Feedman and Lyatkher, 1972). The next step, extension of observed single particle behavior to group particle motion, also presents significant difficulties.

2.3.1.2 Models describing two-phase flow as motion of multi-body system. These models are only of theoretical interest because of the principal possibility to solve the problem of strong, irregular relative mass transport in physically rigorous terms (Levich and Myasnikov, 1966; D'yachenko, 1965). The line of research suggested by these models cannot be realized in the near future as the memory capacity and performance of modern computers do not permit solving problems of this kind. In the near future, this limitation may be overcome with advances in computer design and numerical analyses.

2.3.1.3 Two-phase flow models based on equations of continuous media mechanics. Practical interests call for rapid characterization of detrital mass transport parameters and, primarily, of the regularities of the averaged vertical distribution of suspended sediment. When constructing theoretical models of suspended sediment mass transport by water flows, investigators have to face a number of difficulties:

1. When heavy suspended particles are present in a flow, the problem of turbulence, far from being solved, becomes much more difficult. Complete theoretical description of such flows is absent, while experimental data on the transport of heavy particles are scanty.
2. Full description of a single solid body motion in a viscous liquid possessing an arbitrary velocity field, simultaneously allowing for velocity field action on body movement, and the inverse action of the body on the liquid velocity field is still lacking.
3. There is no generally accepted theory of particle-particle and particle-liquid interaction.
4. The motion of particles relative to the liquid makes the flow more turbulent, which may result in additional dissipation of kinetic energy in the turbulent liquid flow, dampening liquid turbulence. The effect of particles on the turbulent structure is not estimated yet.
5. The action of time-variant orbital components of velocity on turbulence in the wave current regime cannot be estimated quantitatively.
6. Due to the fact that specifications of initial and boundary conditions of two-phase flow are random and approximate, only some probable characteristics of temporal and spatial development of this process can be discussed.
7. Continuous water flow erodes the bottom composed by discrete solid particles. Insufficient theoretical development of solid discrete media mechanics does not permit introducing these real conditions into a model.
8. When interacting with the bottom, the flow itself regulates the amount and composition of sediment transported in suspension, but the principle of this self-regulation is not completely studied.

As the above difficulties cannot be overcome yet, it seems impossible to construct a single, physically rigorous theory, valid for suspended flow in general, and to express it in a numerical form suitable for practical calculations. Thus, one has to construct simplified models of sediment transport limited in applicability to given conditions.

One of the possible ways to construct such models is to turn from the flow of discrete solid particles to a representative flow of a fictitious continuum. Due to the smoothing of phase characteristics by some local volume or by probability, solid and liquid phases are represented in models separately, in the form of continual media, for which equations are written down as for continuous medium flow.

Among these representations, the basic equation system of suspended flow suggested by Frankl (1953, 1955a,b) is the most well-known. This system generalizes Reynolds equations for turbulent flows. It was derived rigorously on the basis of the general equations of continuous medium motion without any additional hypotheses. In Frankl's equations, phase characteristics are averaged by the area occupied by the given phase in some fixed "four-dimensional cylinder" with an elementary volume as a base and a time interval of averaging as an element. Frankl's equations serve as a theoretical basis for a rigorous general definition of dynamic interactions in suspended flow. The same group of theoretical solutions with spatial or spatial-temporal averaging of phase motion parameters includes the equation systems of Slezkin (1952), Dement'ev (1963, 1964), Feedman (1965), Dyunin (1961), and Deemter and Laan (1961).

Models based on a statistic averaging method similar to that adopted in the modern theory of turbulence, are of special interest. This method was used by Buevich (1969), Bazilevich (1969), Nguen-An-Nien (1971), and Crill (1969, 1973). Models of these authors are more simple in terms of the ratio between the number of equations and the number of unknowns, and more universal than the methods of spatial and temporal averaging.

All the above models of two-component flows represent a system of open equations that can be closed only approximately, by the introduction of assumptions decreasing the rigor of the written system and limiting the area of its application.

The equations obtained by closure are complex and are subject to integration only in the simplest particular cases (stationary, flat uniform motion). Solution of these equations is further complicated by difficulties in the specification of boundary conditions that would truly depict the peculiarities of suspended flow motion. To close the system, new theoretical representations are needed. These representations can be based only on abundant, physically valid data from various experiments, which are not available now. Thus, the above approaches to definition of suspended flow motion cannot be used yet for the problem of solid particle distribution in a flow. They actually form a basis for further theoretical studies.

2.3.1.4 Two-phase flow. Let us turn to models presenting suspended sediment transport as a two-phase flow. It is assumed in these models that, theoretically, both phases are present in each point of geometric space; suspended flow is considered to be a nonuniform medium, the density of which varies with the suspension concentration.

Within this concept, the gravity theory of Velikanov (1955) presents an historic interest as he suggested to take account of the flow energy spent on suspended particle transport. More rigorous variants of this theory, in which "suspension work" was allowed for in the equation of energy balance, were developed by Barenblatt (1953, 1955), Kolmogorov (1942), Sinelschikov (1963), and Natishwili (1970). But these models have not been perfected to the degree of practical needs.

For practical purposes, the diffusion theory of sediment motion (Schmidt, 1925; Makkaveev, 1931) is most widely used now. Its basic equations for a uniform material with settling velocity ω_s has the form:

$$\frac{\partial S}{\partial t} + \frac{\partial}{\partial x}\left[\overline{S}\,\overline{U} + \overline{S'U'}\right] + \frac{\partial}{\partial y}\left[\overline{S}\,\overline{V} + \overline{S'V'}\right] + \frac{\partial}{\partial z}\left[\overline{S}(\overline{W} - \omega_s) + \overline{S'W'}\right] = 0 \qquad (2.35)$$

where (x, y, z) are axes of a rectangular coordinate system (z is oriented upward from the bottom, x and y are along and normal to flow), $\overline{U}, \overline{V}, \overline{W}$ are mean values, and U', V', W' are pulsation components of velocity along the x, y, z axes; $\overline{S}$ and S' are mean and pulsational components of suspension concentration. In the above equation, ω_s is the settling velocity of a particle. In the case of heterogeneous sediments, equation (2.35) is used for calculation of narrow fractions, and then the results are summed up in accordance with the fraction input in the suspension composition. By diffusion theory, the turbulent members of (2.35) are parameterized by the coefficient of turbulent diffusion of particles:

$$\overline{S'U'} = -\varepsilon_{sx}\frac{\partial \overline{S}}{\partial x} \qquad (2.36)$$

$$\overline{S'V'} = -\varepsilon_{sy}\frac{\partial \overline{S}}{\partial y} \qquad (2.37)$$

$$\overline{S'W'} = -\varepsilon_{sz}\frac{\partial \overline{S}}{\partial z} \qquad (2.38)$$

Allowing for $\overline{W} = 0$ and substituting (2.36)-(2.38) into (2.35), we get:

$$\frac{\partial \overline{S}}{\partial t} - \frac{\partial}{\partial x}\left[\overline{S}\,\overline{U} - \varepsilon_{sx}\frac{\partial \overline{S}}{\partial x}\right] + \frac{\partial}{\partial y}\left[\overline{S}\,\overline{V} - \varepsilon_{sy}\frac{\partial \overline{S}}{\partial y}\right]$$
$$+ \frac{\partial}{\partial z}\left[\overline{S}\omega_s - \varepsilon_{sz}\frac{\partial \overline{S}}{\partial z}\right] = 0 \qquad (2.39)$$

This equation is applicable to flow of any type and allows for variation of local suspended sediment flows in all three directions. Such an account is necessary for the definition and calculation of the near-bottom suspension field if ripples are present. If ripples are two-dimensional, suspension flow variation along y can be neglected. In a general case of a rippled bottom, variations along x and y can be neglected at a distance $z > \lambda_r$, and for this flow zone the equation is simplified considerably:

$$\frac{\partial \overline{S}}{\partial t} + \frac{\partial}{\partial z}\left[\overline{S}\omega_s - \varepsilon_{sz}\frac{\partial \overline{S}}{\partial z}\right] = 0 \qquad (2.40)$$

The same equation holds true for a flat bottom in the upper smooth phase of sediment motion. For steady-state conditions, if suspension flow is not crossing the upper boundary of flow, equation (2.40) takes the form:

$$\overline{S}\omega_s - \varepsilon_{sz}\frac{\partial \overline{S}}{\partial z} = 0 \qquad (2.41)$$

This equation is commonly used for practical purposes.

2.3.2 Vertical distribution of concentration in steady flow

To understand the main difficulties in concentration calculations, let us consider the simplest case of stationary flow above a flat bottom, defined by equation (2.41). To solve this equation, one should know the functional form of ε_{sz} and the boundary condition close to the bottom, serving as a suspension source. Nearly all of the models assume ε_{sz} to be proportional to the coefficient of water turbulent viscosity:

$$\varepsilon_{sz} = \beta v_T, \qquad (2.42)$$

where β is a coefficient allowing for the inertial properties of particles. Inserting equation (2.42) into equation (2.41) yields the solution:

$$\frac{\overline{S}}{\overline{S}_e} = \exp\left\{-\int_c^z \frac{\omega_s}{\beta v_T}dz\right\} \qquad (2.43)$$

At present, many formulas are based on (2.43), but determination of v_T implies various approaches, the concrete form and analysis of which can be found in Antsyferov and Kos'yan (1986). Generalization of the complete range of approaches to determination of v_T suggests that four types of turbulent viscosity distribution versus height are generally considered:

constant $\qquad\qquad v_T = kU_*h \qquad c \leq z \leq h$ $\qquad\qquad$ (2.44)

linear $\qquad\qquad v_T = kU_*z \qquad c \leq z \leq h$ $\qquad\qquad$ (2.45)

parabolic $\qquad\qquad v_T = kU_*z(1-z/h) \qquad c \leq z \leq h$ $\qquad$ (2.46)

parabolic-constant $v_T = \begin{cases} kU_*z(1-z/h) & c \leq z \leq 0.5z/h \\ 0.25kU_*h & 0.5z/h \leq z < h \end{cases}$ $\quad$ (2.47), (2.48)

where U_* is shear velocity and h is flow depth.

A parabolic distribution based on the physically valid decrease of Reynolds stress from the bottom toward the surface in open flows is most frequently used. In this case the distribution of $\overline{S}(z)$ takes the form:

$$\frac{\overline{S}}{\overline{S}_c} = \left(\frac{h-z}{z}\frac{c}{h-c}\right)^{\omega_s/\beta kU_*}$$ (2.49)

A linear distribution of v_T is commonly used for the modeling of suspended sediment in the bottom boundary layer, where a near-bottom layer of constant Reynolds stresses can be separated. The limitation of (2.49) is that the suspension concentration tends to zero on a flow free surface. For this reason, a parabolic-constant distribution of v_T is sometimes used for the interpretation of experimental data (Coleman, 1970). Comparison of concentration curves for various v_T distributions (Dyer and Soulsby, 1988) showed that in these cases, the relative distribution of concentration varied insignificantly.

Antsyferov and Kos'yan (1986) compared experimental data with formulas derived by various authors. This comparison, however, cannot be reliable without estimation of possible measurement and calculation errors, as the available methods provide for 25% accuracy of U_* determination. As a result, if U_* is measured at the level of bottom roughness elements, mid-flow concentrations, calculated by (2.49), give a three-fold difference (Van Rijn, 1986). Such a range of concentrations, resulting from errors in U_* measurements, is higher than actual scatter given by the suggested models.

It is evident from (2.49) that to calculate concentrations even in the simplest case, β and $\overline{S}_c$ should be known. The various opinions on possible β values can all be divided into three groups:

1. Suspended particles follow all the pulsations of liquid, so β should be analogous to the Schmidt number used for the definition of passive impurity

diffusion (temperature, salts, etc.) in turbulent flows, and which is assumed to be equal to 1.35 from analogy with a neutrally stratified atmospheric boundary layer (Smith and McLean, 1977). The possibility of this evaluation is proved experimentally (Soulsby, Salkied, and Le Good, 1984).

2. Due to sluggishness, suspended solid particles lag behind liquid flow, so β < 1 is assumed.

3. Because of centrifugal forces, heavier particles in suspension, being transported along eddy structures, lead liquid flow, so β > 1 is adopted.

Numerous examples of experimental verification of opinions 2) and 3) are given in Graf (1984) and Antsyferov and Kos'yan (1986) monographies. Analyses of recent publications have not revealed a common opinion. Changing experimental conditions, Farber (1986) got 0.1 < β < 0.5, while in field conditions Lees (1981) got 1 < β < 10. In a Van Rijn study based on experimental data (Van Rijn, 1986) the relation between β and ω_s/U_* is suggested:

$$\beta = 1 + 2\left(\frac{\omega_s}{U_*}\right)^2 \quad \text{for} \quad 0.1 < \frac{\omega_s}{U_*} < 1 \qquad (2.50)$$

which indicates a dominant effect of centrifugal forces upon suspended sand motion. Having measured the pulsation characteristics of suspended sediment concentration and flow velocity in field conditions, Soulsby with co workers (Soulsby et al., 1986) got the value of $\beta = 0.2$, which is in good agreement with the results of Farber's laboratory measurements (Farber, 1986), but differs significantly from Lees and Van Rijn results (Lees, 1981; Van Rijn, 1986). By these authors, such small β values are caused by stable near-bottom stratification of suspended flow, which lowers the turbulence level in comparison to pure water flow. Soulsby et al. (1986) were the first who estimated β by direct determinations of ε_{sz} and v_T from pulsation characteristics. This estimate is physically more valid than an indirect estimate based on concentration curves and used in other works. An estimate of β from concentration profiles always bears some uncertainty caused by the necessity of selection (or fitting) of v_T in accordance with (2.44)-(2.48) for description of experimental data, by the impossibility of estimating the stratification effect on flow parameters, ω_s, U_*, and k in particular, by the variation of the suspension composition in flow measure above heterogeneous bottom sand, and by concentration measuring methods. This seems to be the reason for the absence of a common opinion on the selection of β value for various conditions, so for numerical calculations of suspended sediment concentration in practice $\beta = 1$ is generally used.

The value of the absolute concentration of suspended sediment $\overline{S}_c$ is also vague. The choice of a reference horizon concentration, c, presents a problem.

Usually, a horizon with zero flow velocity is considered as such. For calculations, a method suggested by Smith and McLean (1977) is commonly used,

$$S_c = \frac{S_B \gamma_s c'}{1 + \gamma_s c'} \qquad c' = \frac{\tau_B - \tau_{cr}}{\tau_{cr}} \qquad (2.51)$$

$$c = \frac{\overline{d}}{30} + \alpha_0 \frac{\tau_B - \tau_{CY}}{(\rho_s - \rho)g} \qquad (2.52)$$

in which S_B is sand concentration at the bottom.

River measurements gave the values of $\alpha_o = 26.3$ and $\gamma_s = 2.4 \times 10^{-3}$. This aspect was studied thoroughly in laboratory conditions (Hill, Nowell, and Jumars, 1988) and $\gamma_s = 1.3 \times 10^{-4}$ was obtained. Evaluations of γ_s in marine conditions are still few. Measurements in the intertidal zone (Dyer, 1980) with low values of τ_B gave α_o values similar to those suggested above, while with high τ_B values they differed significantly: $\alpha_o = 100.311$, $\gamma_s = 4.5 \times 10^{-5}$. In continental shelf studies, with a bottom affected by waves and currents, $\gamma_s = 1.6 \times 10^{-5}$ (Wiberg and Smith, 1983), $\gamma_s = 5 \times 10^{-4}$ and $\gamma_s = 2 \times 10^{-5}$ (Drake and Cacchione, 1989) were obtained, being approximately similar to the above values. The variance of the order of magnitude is caused not only by vague physics of the process of sand suspension from the bottom, but also by the methods of evaluation of these parameters, as their selection is still debated.

2.3.3 Vertical distribution of concentration in wave flow

In contrast to steady flow, wave flow is unsteady, which causes additional difficulties in theoretical definition and experimental studies of suspended sediment. Suspension distribution in wave flow is defined by equations (2.39) and (2.40) in accordance with conditions along the x and y axes. Temporal variations of concentration are defined by averaging the time scale, selected for a particular problem solution. If this scale is significantly shorter than the wave period, concentration profiles in various wave phases can be defined by (2.39) and (2.40), while at an averaging scale similar to wave period, concentration changes can be studied during processes that have periods much greater than those of waves (e.g., infragravity waves or periodic storms). In the simplest case of monochromatic and periodic near-bottom water flow, or in a uniform horizontal plane (flat bottom, or at distance greater than the wave length of bedforms), (2.41) can be used for calculation of the period-mean concentration profile needed in practice. Similar to the case of steady flow, the main difficulties arise with the introduction of a coefficient of particle turbulent diffusion and of near-bottom suspension concentration. In principle, to find a value of ε_{sz} from (2.38), one has to know the dependence of $\overline{W'S'}$ and $\overline{S}$ on z, t, but at present we know nothing about the turbulent structure of wave flow with suspension, while a concentration profile is the aim of research. Thus, the assumed proportionality of ε_{sz} to the turbulent viscosity coefficient v_T is used as before. But in contrast to steady flow, physically valid $v_T(z,t)$ relationships in wave flow have not been found yet. The models discussed in Chapter 1 can be used for a few centimeter-thick, near-bottom boundary layer, while above it, with rapid damping of turbulence, orbital water motion can act as a mixing factor. The mechanism of sediment mixing outside the boundary layer is practically unknown, which calls for an indirect approach to determination of ε_{sz}. Equations and models for calculation of wave flow concentration, published in the last 30 years can be divided into three groups. The first one includes studies in which the coefficient of turbulent particle diffusion is based on dimensional theory and generally formalized hypotheses of the physical processes. It has the form:

$$\varepsilon_{sz} = A_s U(z) l(z) \tag{2.53}$$

where $U(z)$ and $l(z)$ are the velocity and length scales in wave flow, and A_s is a proportionality coefficient. This approach was taken in many studies (Rzhanitsyn, 1952; Bashkirov, 1961; Kos'yan, Pakhomov, and Pykhov, 1978; Homma and Horikawa, 1963; Hom-ma, Horikawa, and Kajima, 1965; Wang and Liang, 1975; Skafel and Krishnappan, 1984).

In the second group of studies, values of ε_{sz} are found empirically from measured profiles of the period-mean concentration, with the help of equation (2.41) (Antsyferov and Kos'yan, 1981; Bhattacharya, 1971; Nielsen, 1979; Beach and Sternberg, 1988). It is evident that with this approach to determination of ε_{sz}, the validity of the result depends mainly on the quality of the experimental data used for the determination of numerical coefficients for the first group of models, or of empirical values for the second group of studies. Recall that (2.39)-(2.41) are valid only for uniform grain size sediment and can be applied correctly only if experimental profiles of concentration satisfy this condition. From this point of view, let us consider the most frequently used results of experimental studies of suspended sediment concentration in laboratory and natural conditions. Das (1971), Bhattacharya (1971), Hom-ma (Hom-ma, Horikawa, and Kajima, 1965), and Nakato et al. (1977) measured concentration profiles of suspended sediment with a turbidity meter, which does not give true concentration values if the sand grains are not uniform. To calibrate the meter, suspended sediment was sampled in the measuring point concentration, the mean diameter $\overline{d}$ was found, and the diameter was correlated with the output voltage. A typical calibration curve for the turbidity meter has the form $\overline{S} = kd\Delta V$ (Kennedy and Locher, 1972), where ΔV is the output voltage, and k is a constant coefficient. When experimental measurements are taken, suspended sediment is not sampled, and the value of $\overline{S}$ is found from the mean diameter of bottom sediment $\overline{d}_B$. If the bottom is composed of heterogeneous material, the mean diameter of suspended sediment decreases with distance from the bottom, which was proved experimentally for wave, steady, and complex flows (Antsyferov et al., 1977; Pykhov, Dachev, and Kos'yan, 1980; Antsyferov and Kos'yan, 1981; Antsyferov, Kos'yan, and Longinov, 1973). In these conditions the true value of the concentration $\overline{S}_m$ at any horizon z above the bottom will differ from the concentration $\overline{S}$, measured with the turbidity meter as shown below:

$$\overline{S}_m = \overline{S}\left[\frac{\overline{d}(z)}{\overline{d}_B}\right] \tag{2.54}$$

It follows from (2.54) that the concentration measured by turbidity meter is higher than the true value since $\overline{d}(z)/\overline{d}_B$ is always less than unity and decreases with distance from the bottom. Correction of this discrepancy seems impossible as with this measuring method the form of the $\overline{d}(z)$ curve remains unknown.

Results of Nielsen's study (Nielsen, 1979) were also obtained in laboratory conditions. In these experiments non-uniform sand was used. The concentration of suspended sediment was measured with a siphon. In this measuring method the suspension is sampled, and with the help of grain-size or granular analysis, concentration profiles can be constructed for narrow fractions, the composition of which can be regarded as uniform. Since such an analysis was not made during the study, ε_{sz} was determined by (2.41) with the use of heterogeneous material concentration profiles.

The results of in situ measurements of suspended sediment concentration during a storm, made with sediment traps and used for theoretical analysis, are given by Wang and Liang (1975), Antsyferov and Kos'yan (1977), and Antsyferov et al. (1977). In the first study, concentration profiles of heterogeneous material are analyzed, while in the latter two concentration profiles of narrow sand fractions, selected by detailed grain-size analysis of suspension samples are discussed. These results, however, have one essential limitation as they give the vertical distribution of suspension concentration, averaged for a storm period during which wave parameters varied in a wide range.

A short review of experimental data shows that they cannot be used for correct analysis based on equations written for uniform sediment and for ε_{sz} determination. Nevertheless, this comparison is being made in all the studies. Without any estimates or motivations, sediment composition is assumed to be close to uniform, so that equation (2.41) allowing for $\overline{\omega}_s$ to be equal to the mean value of the settling velocity of bottom sediment ω_B, is used. Such formal use of the equation for uniform sediment evidently leads to physically unreasonable results. In the second group of studies (Bhattacharya, 1971; Nielsen, 1979), ε_{sz} values are found from (2.41) by measured concentration profiles. Bhattacharya experimentally studied wave flow above a sloping bottom covered by a sand layer with a mean diameter $\overline{d}_B = 0.21$ mm and standard deviation in φ units $\sigma_\varphi = 1.32$. The transformation to millimeter scale means that 16% of the particles were coarser than 0.53 mm and 16% were finer than 0.069 mm, i.e., the bottom material is essentially nonuniform. Measurements were made with a turbidity meter in various points of the profile up to the wave breaking zone. All of the measured concentration profiles approximated (2.41) fairly well at $\overline{\omega}_s = \overline{\omega}_B$ and $\varepsilon_{sz} = \varepsilon_{sc}(z/c)$. Fig. 45 shows the values of $\varepsilon_{sz}/\overline{\omega}_B c$ versus the wave relative height H/h. The reference horizon selected close to the bottom was the same for all the concentration profiles. In experiments 2 and 3, the value of $\varepsilon_{sz}/\overline{\omega}_B c$ decreased with H/h growth, which is in contradiction with physical sense. With an increase of H/h, near-bottom velocities grow, and water impact on an eroded bottom increases resulting in more intensive mixing and ε_{sc} growth. Variation of ε_{sc} with H/h in experiment 1 is also physically unexplainable.

Variation of ε_{sc} with H/h by Nielsen's data (Nielsen, 1979) is shown in Fig. 46. Measurements were taken in a wave flume with a horizontal bottom. During three experiments (Fig. 46) water depth and wave period remained constant (0.4 m and 1.7 s respectively), while the sand coarseness and composition varied. Each experiment included a series of measurements at various wave heights. The vertical distribution of the concentration was described by (2.41), if $\overline{\omega}_s = \overline{\omega}_B$ and $\varepsilon_{sz} = const$ were assumed. Plausible ε_{sc} variations with H/h were observed only in experiment 1, which is quite understandable taking into account the selected sand composition. The degree of sand size non-uniformity is characterized by the sorting coefficient $\sigma_w/\overline{\omega}_B$ where σ_w is the

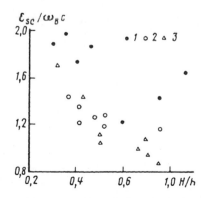

FIGURE 45   Variations of $\varepsilon_{sc}/\overline{\omega}_B$ with H/h, by Bhattacharya (1971) experimental data. 1: $T = 1.5$ s; 2: $T = 1.05$ s; 3: $T = 1.05$ s.

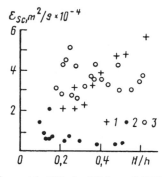

FIGURE 46   Variation of ε_{sc} with H/h, by Nielsen (1979) experimental data. 1) T = 1.7s; ω_B = 0.067m/s; σ_w/ω_B = 0.13; 2) T = 1.7s; ω_B = 0.0054m/s; σ_w/ω_B = 0.70; 3) T = 1.7s; ω_B = 0.021m/s; σ_w/ω_B = 0.39.

standard deviation. As one can see, only the sand used for experiment 1 is close to uniform, so these measurements can be interpreted with the help of (2.41). In other experiments, the sand is evidently nonuniform, and use of (2.41) with $\overline{\omega}_s = \overline{\omega}_B$ for determination of ε_{sc} gives physically invalid results.

In the first group works, the coefficient A_s in (2.53) is also found from a heterogeneous sand concentration profile, and it can be easily shown that formulas derived for ε_{sz} do not reflect their true variations. Let us consider a sand with heterogeneous composition, consisting of a set of narrow fractions. Each fraction will be characterized by a dimensionless value of settling velocity $\omega_* = \omega_i/\overline{\omega}_c$, where ω_i is the settling velocity of the ith fraction and $\overline{\omega}_c$ is the mean settling velocity of heterogeneous sand at the horizon c. Let the normalized weight density of the suspended particle distribution at c by their settling velocity ω_* be $\varphi_c(\omega_*)$. If the suspension and distribution of particles from narrow fractions proceed independently under the same hydrodynamic conditions, then the vertical distribution of the suspension concentration of each fraction is described by (2.41), while for heterogeneous sand it is found by integrating (2.41) with respect to ω_* from 0 to ∞ with account of $\varphi_c(\omega_*)$.

$$\overline{S}_\Sigma(z) = \overline{S}_{\Sigma c}\int_0^\infty \varphi_c(\omega_*)\exp(-N\omega_*)d\omega_* \tag{2.55}$$

where

$$N = \int_c^z \frac{\overline{\omega}_c}{\varepsilon_{sz}}dz \tag{2.56}$$

where $\overline{S}_{\Sigma c}$ is the mean concentration of heterogeneous material at a horizon c.

It follows from (2.55) that for heterogeneous sand, the relationship between ε_{sz} and ω_* should be known. It has been mentioned before that in the first group works ε_{sz} is assumed to be independent of ω_*, and is only a function of z. But as Coleman showed (Coleman, 1970), ε_{sz} values for various fractions in steady flow vary under the same hydrodynamic conditions. This is also confirmed by Antsyferov and Kos'yan (1977) data, so when (2.55) is compared with experimental data, one should specify the assumed form of the ε_{sz} relationship with both z and ω_*. Otherwise the result will be confusing, which can be illustrated by the concentration profiles of heterogeneous sand plotted by Nielsen (1979) and Bhattacharya (1971) and shown in Fig. 47.

On the one hand, under a formal approach they can be approximated by equation (2.41) for uniform material, and if $\omega = \bar{\omega}_c = const$ is assumed, we get:

$$\bar{S}_\Sigma(z) = \bar{S}_{\Sigma c} e^{-N_1} \tag{2.57}$$

where

$$N_1 = \int_c^z \frac{\bar{\omega}_c}{\varepsilon_m(z)} dz \tag{2.58}$$

We will call $\varepsilon_m(z)$ a dummy value of coefficient of turbulent diffusion since experimental data represent the depth distribution of heterogeneous sand. In Nielsen's experiment (Fig. 47a) $\varepsilon_m = const$, while in Bhattacharya's experiment (Fig. 47b) $\varepsilon_m \sim (z/c)$. On the other hand, the concentration profile of heterogeneous sand is described rigorously by (2.55) with the true value of ε_{sz} for each fraction, although its form remains unknown in advance. After the authors of the first group of studies assumed that ε_{sz} does not depend on ω_*, but is a function of z only:

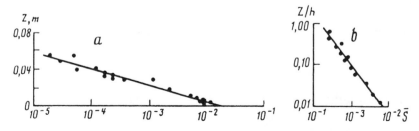

FIGURE 47   Experimental profiles of heterogeneous material concentration. a) $T = 1.7s$; $h = 0.4m$; $H = 0.21m$; $d_B = 0.082mm$ (Nielsen, 1979); b) $T = 1.04s$; $h = 0.2m$; $H = 0.096m$; $d_B = 0.21mm$ (Bhattacharya, 1971).

$$\varepsilon_{sz} = \varepsilon_1(z) \tag{2.59}$$

Nielsen (1979) showed that $\varphi_c(\omega_*)$ for a heterogeneous sand is approximated well by the gamma distribution

$$\varphi_c(\omega_*) = \frac{1/\alpha_*^{1/\alpha_*}}{\Gamma(1/\alpha_*)} \exp(-\omega_*/\alpha_*)\omega_*^{\frac{1}{\alpha_*}-1} \tag{2.60}$$

where $\Gamma(1/\alpha_*)$ is the gamma function; and $\alpha_* = [\sigma_{\omega c}/\omega_c]^2$. Substituting (2.60) into (2.55) and integrating with respect to ω_* going from 0 to ∞ we get

$$\overline{S}_\Sigma(z) = \overline{S}_{\Sigma c}\left[\frac{1}{(1+\alpha_* N)}\right]^{1/\alpha_*} \tag{2.61}$$

Allowing for $e_m = dN_1/dz$ and $e_1 = dN/dz$, from (2.57) and (2.61) we get:

$$\frac{\varepsilon_m}{\varepsilon_1} = 1 + \alpha_* N \tag{2.62}$$

or

$$\varepsilon_1 = \varepsilon_m \exp\left\{-\int_c^z \omega_c \alpha_* dz/\varepsilon_m\right\} \tag{2.63}$$

It follows from (2.63) in particular that at $\varepsilon_m = const$, the value of ε_1 is a function of z and depends on the degree of particle size nonuniformity which is characterized by α_*. Variation of $\varepsilon_m/\varepsilon_1$ with N and α_* is shown in Fig. 48, and it is seen clearly that the true value of ε_1 with $z > c$ is always less than ε_m. This difference is maximal at high values of N and α_*.

Suppose then that ε_{sz} is an arbitrary function of z and depends linearly on ω_*:

$$\varepsilon_{sz} = \omega_* \varepsilon_1(z) \tag{2.64}$$

Then for any form of $\varphi(\omega_*)$, it follows from (2.55) that

$$\overline{S}_\Sigma(z) = \overline{S}_{\Sigma C} \exp\left\{-\int_c^z \omega_c dz/\varepsilon_1(z)\right\} \tag{2.65}$$

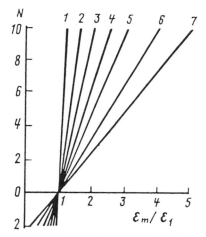

FIGURE 48   The effect of composition heterogeneity on the variation of the coefficient of particle diffusion for heterogeneous material. 1) $a^* = 0.02$; 2) = 0.06; 3) = 0.10; 4) = 0.15; 5) = 0.20; 6) = 0.30; 7) = 0.40.

Comparing (2.65) with the experimental profile of concentration described by (2.57), we get

$$\varepsilon_1(z) = \varepsilon_m(z) \tag{2.66}$$

Thus, the same experimental data described by the formal equation for a uniform material give physically different relationships of ε_{sz} if its functional dependence in (2.63) is specified in advance. It means that real values of ε_{sz} cannot be found from experimental profiles of heterogeneous sand concentration even with the use of a rigorous formula (2.56), and that all the expressions suggested for ε_{sz} in both groups of studies do not present actual variations of the diffusion coefficient in a wave flow. This, in fact, explains the variety of ε_{sz} expressions suggested in studies which, on the one hand, differ from each other considerably but, on the other hand, formally agree well with the experimental data of each author due to the fitting of empirical coefficients. Thus, to determine physically valid values of ε_{sz}, experimental data characterizing the vertical distribution of uniform sediment concentration are needed. To get them, one has to experiment either with a uniform size sand, or with heterogeneous sand with separation of narrow fraction concentration profiles. The first route is practically impossible as the required amount of uniform material could hardly be obtained even for laboratory studies, while in marine conditions sediment is always nonuniform. The second route seems more real at present.

Some symbiosis of these models was reached in the model for calculation of the vertical profile of a period-mean suspension concentration suggested

by Kos'yan and Pakhomov (1981) and Kos'yan (1983b) in which a solid particle diffusion coefficient is determined accounting for their settling velocity.
The resulting formula reflects particle mixing due to turbulence in the bottom boundary layer and due to orbital motion outside it:

$$\bar{S}(z)/\bar{S}_c = \exp\left(-\int_c^z \omega_s \left[\frac{a_1 H^2 \sinh kz}{T \sinh kh} + \frac{a_2(U_m - \omega_s)\dfrac{z}{\delta_w}}{1 + a_3 \dfrac{z}{\delta_w}\exp\dfrac{z}{\delta_w}}\right]^{-1} dz\right) \quad (2.67)$$

where

$$a_1 = \frac{\pi}{2\sqrt{2}}$$

$$a_2 = 116\frac{\rho}{\rho_s - \rho}\left(\frac{\nu^2}{g}\right)^{1/3}$$

$$a_3 = 0.06$$

and δ_w is bottom boundary layer height, found from (1.7). Being tested by
data from several laboratory experiments, this formula demonstrated good
agreement with their data and is recommended for the calculation of concentration profiles of separate fractions under monochromatic wave conditions.
The above simple models are generally based on a formal approach to the
determination of the coefficient of turbulent diffusion.

Models of the third group are more promising as a turbulent closure
model is applied for a two-phase flow. With reference to suspended sediment
motion in the wave boundary layer, such models are still very few (Justesen
and Fredsøe, 1985; Hagatun and Eidsvik, 1986; Justesen, 1988). They are
based on equations of momentum preservation, turbulent kinetic energy
transport, and turbulent energy dissipation transport for a wave boundary
layer (see Chapter 1). In our case, the equation of mass transport in the form
of (2.40) is used, while the equation of turbulent energy transport and energy
dissipation transport is enriched with terms allowing for turbulent energy
dissipation on suspended particle transport. Figs. 49 and 50 illustrate calculations and comparisons with experimental data (Hagatun and Eidsvik, 1986)
and clearly show variations of the main parameters of two-phase wave flow in
$(z/a_m, \omega t)$ coordinates.

Description of suspended sediment motion in the wave breaking zone is
more difficult, as the model for a wave boundary layer is insufficient due to

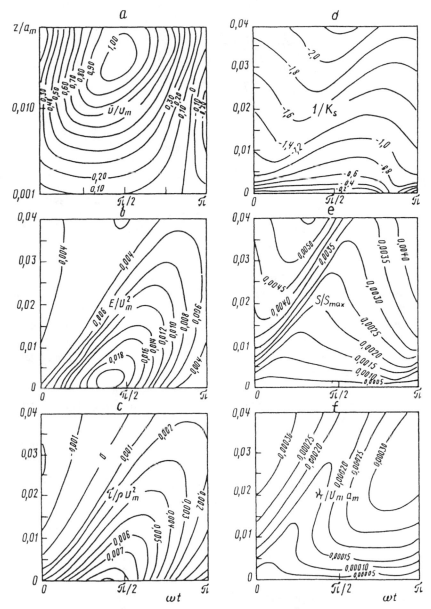

FIGURE 49   Theoretically calculated distribution of (a) amplitudes (U/U_m), (b) turbulent kinetic energy (E/U_m^2), (c) Reynolds stress ($\tau/\rho U_m^2$), (d) length scale (l/K_s), (e) suspension concentration ($\bar{S}/S_{max}$), and (f) turbulent viscosity ($\nu_T/U_m a_m$) in plane ωt, z/a_m plane. Model parameters: $U_m = 1.0$ m/s; $T = 10$ s; $S_{max} = 0.3$; $\bar{d} = 0.2$ mm; $K_s = 0.001$ m (Hagatun and Eidsvik, 1986).

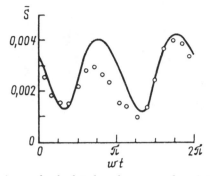

FIGURE 50   Comparison of calculated and measured variations of concentration during a wave period at $z = 1.8$ cm in a bottom boundary layer (experiment 5, of Staub, Swendsen, and Johnsson, 1983): $T = 9.1$ s; $U_m = 1.21$ m/s; $\bar{d} = 0.19$ mm (Hagatun and Eidsvik, 1986).

the generation of the major part of turbulent energy in the upper flow during wave breaking. An attempt was made to allow for turbulent energy generation of wave breaking and to account for it in the equation of turbulent energy transport (Deigaard, Fredsøe, and Hadegaard, 1986). In this study, Fredsøe's model is used (Fredsøe, 1984) to which the equation of turbulent energy transport is added with a modified term standing for turbulent energy generation. Theoretical calculations by this model show that the form of the velocity profile depends on the H/h ratio in the wave breaking zone (Fig. 51) and on the gT^2/h parameter. Comparison of calculated and measured distribu-

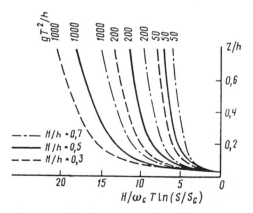

FIGURE 51   Theoretical profiles of concentration within the breaking zone (Deigaard, Fredsøe, and Hadegaard, 1986).

tions of concentration in laboratory conditions with wave breaking by spilling gives good agreement (Fig. 52).

For numerical calculations by all the models, the reference concentration $\bar{S}_c$ should be known. A common approach to determination of $\bar{S}_c$ and choice of reference horizon c is lacking, as in the above cases for steady flows. The near-bottom concentration is usually assumed to be proportional to the amplitude of the near-bottom orbital velocity U_m or to the bottom shear stress τ_B. Suggested formulas differ from each other significantly. For example, Vincent, Young, and Swift (1982) got $\bar{S}_c \approx U_m^2$ at $c = 1$ cm above bottom in field conditions, while in laboratory experiments Nielsen (1986) found $\bar{S}_c \approx U_m^6$ by extrapolation of concentration profiles toward the bottom. The linear relationship between the concentration and orbital velocity obtained by near-shore measurements with $c = 1$ cm above the bottom is suggested by Lesht, Clarke, and Young (1980) and Clarke, Lesht, and Young (1982). The marked

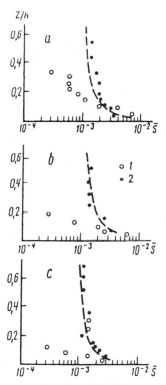

FIGURE 52 Comparison of calculated profiles of suspended sediment concentration with laboratory data (Deigaard, Fredsøe, and Hadegaard, 1986). a) $h = 0.28$ m; $H = 0.1$ m; $T = 1.4$ s; b) $h = 0.34$m; $H = 0.12$ m; $T = 1.4$ s; c) $h = 0.38$ m; $H = 0.13$ m; $T = 1.4$ s; 1: unbroken wave; 2: broken wave.

difference in power indices of U_m shows definitely all the difficulties of $\bar{S}_c$ prediction and is caused by the absence of clear physical knowledge of sediment suspension by waves above bottoms of various types. Among the formulas of this type one should be noted especially (Kos'yan, 1986):

$$\bar{S}_c = 1.3 \cdot 10^{-6} \left(\frac{U_m}{\omega_s} \right)^{5/4} \tag{2.68}$$

derived from measuring results in the Black Sea and Mediterranean testing sites. In (2.68) the reference horizon is $c = 30$ cm above the bottom, and ω_s values are also measured at this horizon.

In recent years, when temporal variations of concentration have been studied for infragravity waves (Beach and Sternberg, 1988) or for longer time periods, equations (2.51) and (2.52) have been used, in which τ_B is taken as absolute value and is time-variant. Values of γ_s are either taken from studies of steady flows cited above, or are estimated by in situ measurements. Thus, by extrapolation of measured concentration profiles, Beach and Sternberg (1988) obtained $\gamma_s = 1.05 \times 10^{-4}$ for $c = 0.1$ cm. Comparable values of $\gamma_s = 1.5 \times 10^{-4}$ and $\gamma_s = 4.5 \times 10^{-4}$ were obtained by Shi, Larsen, and Downing (1985).

This brief review of approaches to determination of $\bar{S}_c$ clearly shows that this problem has not been solved yet, and for practical estimates experimental data are desirable to test the suggested formulas and correct them for the approximate region. For this reason (2.68) is used in Chapter 7 for the longshore sediment transport estimates for the Bulgarian Black Sea coast as it was based partly on Kamchia site data.

It should be also noted that data on the near-bottom concentration in the wave-breaking zone is very limited. Available data (Nielsen, 1979; Staub, Swendsen, and Johnsson, 1983) show that $\bar{S}_c$ values for unbroken wave and for the zone of wave breaking by spilling are nearly the same. This is clearly seen in Fig. 52, where Staub's (Staub, Swendsen, and Johnsson, 1983) concentration profiles are shown for breaking (solid circles) and unbroken waves at similar heights and periods. Thus, it becomes evident that the near-bottom suspension concentration is controlled by turbulence of the bottom boundary layer.

2.3.4 The calculation of heterogeneous sand concentration in a wave flow. Let the reference density of such sand distribution by the settling velocity ω_s at any above bottom horizon z be $\varphi(\omega_s)$. Then the concentration of the suspension sand fraction with the settling velocity ω_s in the interval $d\omega_s$ and the concentration of suspension heterogeneous sand are related as

$$\bar{S}(z) = \varphi(\omega_s) d\omega_s \bar{S}_\Sigma(z) \tag{2.69}$$

From (2.41) the concentration of each sand fraction is described by the equation

$$\overline{S}(z)\omega_s + \varepsilon_{sz}\left(\frac{\partial \overline{S}}{\partial z}\right) = 0 \tag{2.70}$$

Substituting (2.69) into (2.70) and integrating with respect to ω_s from 0 to ∞, we get

$$\overline{S}_\Sigma(z)\overline{\omega}_\Sigma(z) + \overline{\varepsilon}_\Sigma(z)\frac{\partial \overline{S}_\Sigma(z)}{\partial z} = 0 \tag{2.71}$$

where

$$\overline{\omega}_\Sigma(z) = \overline{\omega} + \int_0^\infty \varepsilon_{sz}\frac{\partial \varphi(\omega)}{\partial z}d\omega; \quad \overline{\omega} = \int_0^\infty \omega\varphi(\omega)d\omega; \quad \overline{\varepsilon}_\Sigma = \int_0^\infty \varepsilon_{sz}\varphi(\omega)d\omega$$

In these formulae $\overline{\omega}_\Sigma$ and $\overline{\varepsilon}_\Sigma$ are mean values of the settling velocity and solid particle diffusion coefficient for heterogeneous sand at horizon z. The solution of (2.71) with $\overline{S}_\Sigma(z-c) = \overline{S}_{\Sigma c}$ will take the form of

$$S_\Sigma(z) = S_{\Sigma c}\exp\left[-\int_c^z (\overline{\omega}_\Sigma/\overline{\varepsilon}_\Sigma)dz\right] \tag{2.72}$$

Thus, comparing (2.71) with experimental data, we can evaluate $\overline{\omega}_\Sigma(z)/\overline{\varepsilon}_\Sigma(z)$ for heterogeneous sand. The distribution density of suspension particles at any horizon can be found from measurement results using the measured sand size distribution. If variations along z are neglected, the value of $\overline{\varepsilon}_\Sigma(z)$ can also be evaluated as $\overline{\omega}_\Sigma(z)$ is easily calculated from $\varphi_\Sigma(z)$. The above experimental data cannot be used for evaluation of $\overline{\omega}_\Sigma(z)/\overline{\varepsilon}_\Sigma(z)$ since the value of $\phi(\omega_s)$ is unknown. Detailed measurements with the use of suspended sediment traps were carried out from a trestle down to 4.5 m depth in the two-meter near-bottom layer on 15-20 horizons above the bottom. Characteristics of the bottom slope profile and storm-induced bottom deformations, bottom sediment composition in points of concentration measurements, hydrodynamic conditions, peculiar vertical distribution of suspension concentration and composition in various time intervals during a storm, and variations of the intensity of bottom sediment suspension along the bottom slope in various storm phases are discussed in detail in a number of studies (Nikolov and Pykhov, 1980; Pykhov, Dachev, and Kos'yan, 1980). As wave parameters varied insignificantly during concentration profile measurements, the hydrodynamic regime can be considered as stationary.

A preliminary analysis of suspended sediment concentration profiles showed that ε_{sz} in (2.41) varies in a fashion

$$\varepsilon_{sz} = \varepsilon_{sc}\left(\frac{z}{c}\right)^{\alpha} \tag{2.73}$$

while $\overline{\omega}_{\Sigma}/\overline{\varepsilon}_{\Sigma}$ in (2.72) for heterogeneous sand concentration is

$$\frac{\overline{\omega}_{\Sigma}}{\overline{\varepsilon}_{\Sigma}} = \left(\frac{\overline{\omega}_{\Sigma}}{\overline{\varepsilon}_{\Sigma}}\right)_c\left(\frac{z}{c}\right)^{-\tilde{\alpha}} \tag{2.74}$$

where ε_{sz} and $(\overline{\omega}_{\Sigma}/\overline{\varepsilon}_{\Sigma})$ are the values of these parameters at horizon c. Substituting (2.73) into (2.41) and (2.74) into (2.72) and integrating it, we get for a fraction:

$$\ln\left[\frac{\overline{S}(z)}{\overline{S}_{sc}}\right] = -\left(\frac{\omega_{sc}c}{\varepsilon_{sc}}\right)\ln\left(\frac{z}{c}\right) \quad \text{at} \quad \alpha = 1 \tag{2.75}$$

$$\ln\left[\frac{\overline{S}(z)}{\overline{S}_{sc}}\right] = -\frac{\left(\frac{\omega_{sc}c}{\varepsilon_{sc}}\right)\left[1-\left(\frac{z}{c}\right)^{1-\alpha}\right]}{(\alpha-1)} \quad \text{at} \quad \alpha \neq 1 \tag{2.76}$$

and for heterogeneous sand

$$\ln\left[\frac{\overline{S}_{\Sigma}(z)}{\overline{S}_{\Sigma c}}\right] = -\beta\ln\left(\frac{z}{c}\right) \quad \text{at} \quad \tilde{\alpha} = 1 \tag{2.77}$$

$$\ln\left[\frac{\overline{S}_{\Sigma}(z)}{\overline{S}_{\Sigma c}}\right] = -\frac{\beta[1-(z/c)^{1-\tilde{\alpha}}]}{(\tilde{\alpha}-1)} \quad \text{at} \quad \alpha \neq 1 \tag{2.78}$$

where α, $\tilde{\alpha}$, ε_{sc} and $\beta = (1/c)(\overline{\omega}_{\Sigma}/\overline{\varepsilon}_{\Sigma})_c$ from (2.73)-(2.76) depend on the surface wave parameters, particle settling velocity, and should be derived empirically. In our studies, the lower measured horizon was 0.1-0.2 m above bottom. In principle, any value of c can be chosen. We chose $c = 0.20$ m in order to get an empirical relationship for $\overline{S}_{\Sigma c}$ as close to the bottom as possible.

Suspension samples were analyzed for grain size with the following fractions being separated: <0.1; 0.1-0.125; 0.125-0.15; 0.15-0.25; 0.25-0.315; 0.315-0.50; 0.5-0.8; 0.8-1.0; 1.0-1.25; 1.25-1.6; 1.6-2.0; 2.0-2.5; and >2.5 mm. Thus, each profile of heterogeneous sand concentration corresponded to a series of concentration profiles by fractions.

Optimal values of α, $\tilde{\alpha}$, ε_{sc}, and β for which equations (2.75)-(2.78) approximate measurement data with the least error, were found by the least squares method for each concentration profile. To calculate the error, the standard deviation σ_s of the experimental value of the concentration logarithm $\ln[\bar{S}_\Sigma(z)/\bar{S}_{\Sigma c}]$ or $\ln[\bar{S}(z)/\bar{S}_c]$, from calculated values was first found. Then the error σ_s was calculated from E_s

$$E_s = e^{\sigma_s} \tag{2.79}$$

If experimental data are ideally described by the theoretical curve, then $\sigma_s = 0$ and $E_s = 1$. If the experimental data are scattered relative to the approximating curve, then $E_s > 1$ and the relative error is characterized by the deviation of E_s from unity. Concentration profiles for 50 heterogeneous sand samples were analyzed. Unfortunately, only in some of them was the weight of the suspension samples sufficient for accurate grain-size analysis at all points of measurement in the two-meter layer. In most cases, even at $z > 0.3 \rightarrow 0.8$ m from the bottom, the weight of the suspension sample did not exceed a few grams. In this situation, weight of the sample for grain-size analysis varies at different profile points. As a result, the composition analysis error grows in the upper horizons where the sample weight is fractions of a gram and quantitative analysis of the concentration profiles of separate fractions becomes senseless. Thus, to estimate the fractional values of α and ε_{sc}, two cases were chosen in which the sample weight was at least 20 g at each measuring point. In Table 1 α and ε_{sc} values are given for coarse grained sand as a bottom material. The

TABLE 1 α and ε_{sc} per fraction values for coarse-grained sand

Fraction mm	$\bar{\omega}_s$ m/s	P_i %	$\varepsilon_{sc} \cdot 10^{-2}$ m²/s	α	E_s
0.8–1.0	0.105	2.5	1.62	0.4	1.13
0.5–0.8	0.092	27.4	156	0.3	1.20
0.315–0.5	0.061	42.4	101	0.5	1.17
0.25–0.315	0.040	9.1	61	0.85	1.14
0.15–0.25	0.022	9.5	47	1.15	1.17
0.125–0.15	0.014	6.6	39	1.30	1.13
0.10–0.125	0.011	2.5	34	1.30	1.19

measuring site depth was $h = 2.4$ m, the wave mean height $\overline{H} = 0.61$ m, and the mean period $\overline{T} = 5.15$ s. Similar data for medium-grained sand are given in Table 2 ($h = 3.82$ m, $\overline{H} = 1.09$, $\overline{T} = 4.8$ s).

Settling velocity approximation errors, E_s, and the weight percent of each fraction are also given in the Tables. These values were found from the mean diameter of each fraction and from the curve of sand particle size versus settling velocity, given in Swart (1976). In both cases ε_{sc} decreased and increased with a decrease of settling velocity. These results show that the value of diffusion coefficient for separate narrow fractions of sand and its variance along z depend on ω_s considerably, which was not accounted for in the above studies. Parameters characterizing the concentration profiles of a polyfractional material were estimated for these two cases as $\tilde{\alpha} = 0.75$; $\overline{\varepsilon}_{\Sigma c} = 1.04 \times 10^{-2}$ m²/s, $E_s = 1.12$ for coarse-grained sand; and $\tilde{\alpha} = 1.5$; $\overline{\varepsilon}_{\Sigma c} = 3.0 \times 10^{-3}$ m²/s, $E_s = 1.05$ for medium-grained sand. Formally, the $\tilde{\alpha}$ and $\overline{\varepsilon}_{\Sigma c}$ values are close to α and ε_{sc} values for the dominating suspension fraction. ω_{Σ} values estimated by these data for separate profile points are presented in Tables 3 and 4.

Mean values of the standard deviations σ_ω for settling velocity $\overline{\omega}_{\Sigma}(z)$ for a heterogeneous suspension are also given. It can be easily seen that $\overline{\omega}_{\Sigma}(z)$ differs between coarse- and medium-grained sands from $\overline{\omega}_{\Sigma}(z)$ within $\approx 10\%$. This means that $\phi(\omega_s)$ varies insignificantly with increasing z. Two conclusions can be drawn from the above estimates:

1. Equation (2.41) holds true for the definition of time-average concentration. In this case the coefficient of turbulent particle diffusion for separate fractions depends on ω_s, and this fact was ignored in previous models.
2. Equation (2.41) cannot be used for estimates of ε_{sz}, as the equality of suspension flows, on which (2.41) is based, was not followed during these measurements.

To solve this dilemma, profile measurements alone are not enough; data on diffusive flow of particles eroded from bottom, $\overline{W'S'}$ are needed, which could

TABLE 2 α and ε_{sc} per fraction values for medium-grained sand

Fraction mm	$\overline{\omega}_s$ m/s	P_i %	$\varepsilon_c \cdot 10^{-2}$ m²/s	α	E_s
0.5–0.8	0.092	2.1	104	1.05	1.32
0.315–0.5	0.061	13.0	66	1.10	1.25
0.25–0.315	0.040	12.1	36	1.55	1.13
0.15–0.25	0.022	39.8	25	1.60	1.09
0.125–0.15	0.014	25.7	16	1.60	1.07
0.10–0.125	0.011	7.3	18	1.25	1.04

TABLE 3   The suspension composition and ϖ values on various above bottom horizons for coarse-grained bottom sand

z m	$\overline{\omega}$ m/s	$\sigma_{\overline{\omega}}$ m/s	$\sigma_{\overline{\omega}}/\omega$	$\overline{\omega_\Sigma}$ m/s	$\overline{\omega_\Sigma}/\omega$
0.20	0.061	0.027	0.44	0.060	0.98
0.48	0.055	0.028	0.51	0.047	0.86
0.79	0.046	0.029	0.63	0.045	0.98

TABLE 4   The suspension composition and ϖ values on various above bottom horizons for medium-grained bottom sands

z m	$\overline{\omega}$ m/s	$\sigma_{\overline{\omega}}$ m/s	$\sigma_{\overline{\omega}}/\omega$	$\overline{\omega_\Sigma}$ m/s	$\overline{\omega_\Sigma}/\omega$
0.19	0.028	0.018	0.64	0.026	0.93
0.48	0.026	0.017	0.65	0.024	0.92
0.88	0.024	0.016	0.66	0.023	0.96
1.29	0.023	0.014	0.61	0.023	1.00
1.90	0.022	0.012	0.65	0.023	1.04

be obtained only from measured pulsation parameters of suspension flow (see example at the end of this chapter).

If the validity of (2.41) for practical estimates is recognized, empirical values of $\tilde{\alpha}$, b, and $\overline{S}_{\Sigma c}$ could be found from profile measurements and used for calculations with (2.77) and (2.78). It should be noted as a conclusion that the exact physics of sediment transport processes remains vague, which affects modeling even of the simple situations discussed above. Knowledge of "bursting" processes of turbulent energy generation in the boundary layer seems most promising for these cases. Intensive simultaneous application of visual and instrumental methods for studies of laboratory boundary layer turbulence showed that near-bottom generation of turbulence occurs as a "bursting" process. It is characterized by quasi-periodic fluid ejections upward from the bottom and an equally energetic fluid inrush toward the bottom (Cantwell, 1981). Elements of this process can be traced on simultaneous records of U', W', and $U'W'$ values. Four typical elements are recognized from these records:

1. Ejection of slower fluid outward from the bottom into the flow body:

$$U' < 0, \ W' > 0, \ RS_1 = -U'W' > 0$$

2. Inrush of faster fluid elements into the bottom zone of slower velocities:

$$U' > 0, \ W' < 0, \ RS_2 = -U'W' > 0$$

3. Outward interaction of fluid elements:

$$U' > 0, \ W' > 0, \ RS_3 = -U'W' < 0$$

4. Inward interaction of fluid elements:

$$U' < 0, \ W' < 0, \ RS_4 = -U'W' < 0$$

where RS_i are instantaneous values of the Reynolds stresses corresponding to these elements. The record-average Reynolds stress will be: $\overline{RS_1 + RS_2 + RS_3 + RS_4}$.

Study of the time series of the instantaneous values of velocity and Reynolds stresses can indicate participation of these elements in the resulting Reynolds stress which controls flow dynamics and, consequently, sediment transport. In situ studies of these phenomena in marine bottom boundary layers have just begun with first results already published for the upper shelf (Gordon, 1974; Heathershaw, 1974, 1976; Gordon and Witting, 1977; Soulsby, 1983). Fig. 53 shows a typical time series of Reynolds stress instantaneous values measured at a 2 m depth in a tidal estuary (Gordon and Witting, 1977). RS_1 are shaded and RS_2 are dotted. It is clearly seen that fluid ejections and sweeps control 30% of the Reynolds stress averaged over the measurement time. The duration and time interval of ejections and inrushes control the scales of boundary layer eddy structures. An analysis made by the same authors showed that the duration varied from 2 to 20 s with a mean value of 9 s. Instantaneous Reynolds stresses were 3 to 5 times greater than the record-

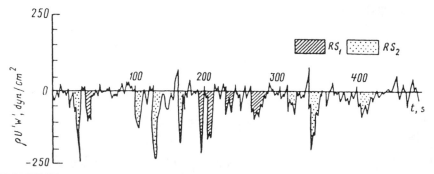

FIGURE 53 Typical time-series of instantaneous values of Reynolds stress (Gordon and Witting, 1977).

mean value. This conclusion is very important as sediment transport is defined by the instantaneous flow conditions. Modeling, unfortunately, is usually based on average parameters.

Near-bottom marine measurements (Soulsby, Atkins, and Salkield, 1987; Heathershaw and Thorn, 1985) show that the suspension of bottom sediment correlates well with fluid ejections while the bottom sediment transport is related to the sweeps. The ejection significance in sediment suspension is illustrated by Fig. 54 which shows good agreement between the velocity of particle outward motion and ejected fluid velocity.

Studies of two-phase flow pulsation parameters (Soulsby, Atkins, and Salkield, 1987; Soulsby, Salkield, and Le Good, 1987) can also give new insight into the physical aspect of the phenomena. Examples of such measurements are shown in Fig. 55. These measurements were taken in a tidal estuary at 13 and 33 cm above a sand-covered bottom with particle mean diameter of 0.165 mm. The instantaneous concentration was registered by a pulse counting device, and velocity was measured by electromagnetic velocity meters at horizon close to a 25 m long and 0.75 m high sand wave crest. As one would expect, the suspension specific discharge caused by turbulent diffusion $\overline{U'S'}$ is

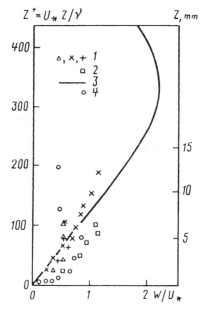

FIGURE 54   Experimental data on vertical velocity of particles moving upwards in the bottom boundary layer (Summer, 1986). 1: Summer and Deigaard, 1981; 2: Summer and Oguz, 1978; 3: Grass, 1974; 4: Brodkey, Wallace, and Eckelmann, 1974; Summer, 1986.

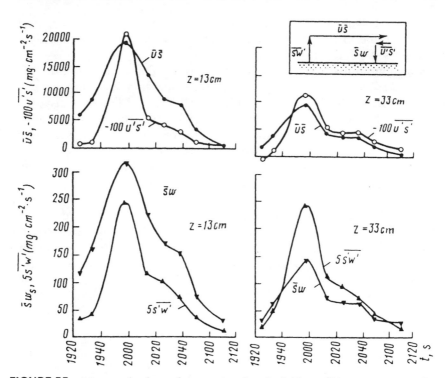

FIGURE 55   Measured values of suspension flow in field conditions at $z = 13$ and 33 cm above bottom (Soulsby, 1986).

negative and nearly 100 times less than the average $\overline{US}$, which is clearly seen. Diffusive suspension flow caused by erosion, $\overline{W'S'} > 0$, is nearly constant and is 4 to 6 times higher than suspension flow, $\overline{S}\omega_S$, caused by particle settling at these horizons. This means that suspension flows are not balanced at each vertical point, as is assumed for modeling. As $\overline{U'S'} < 0$ and $\overline{S'W'} > 0$, concentration growth at a given horizon $S' > 0$ is possible at $U' < 0$ and $W' > 0$ which control fluid ejection, while its decrease $S' < 0$ occurs at $U' > 0$ and $W' < 0$, which control the fluid inrush toward the bottom. This proves the immense role of the "bursting" process in the formation of a suspension concentration field and also shows that our present knowledge of physics of this phenomena is very poor, while further progress will depend on experimental research of the turbulent structure of suspension flows.

PART

II

Field Research

CHAPTER
3

Measuring Methods

The dynamic processes of the nearshore zone are extremely complex and modeling these processes depends greatly on the quality of in situ experimental studies. Surface waves, nearshore currents, and sediment transport control the long-term underwater slope and shoreline modifications and are being studied intensively worldwide in natural conditions in order to solve both fundamental and applied tasks. Since one-of-a-kind instruments are generally used in most studies, strict instrument control is not possible and the results are hard to compare. The great significance of nearshore measurement methodology is well recognized and has gained due attention in several texts on nearshore processes (Antsyferov and Kos'yan, 1986; Coastal sediments, 1987; Nearshore Dynamics and Coastal Processes (NDCP), 1988; Basinski, 1989; Aibulatov, 1990). One of them (Basinski, 1989) deserves special attention as it is completely devoted to a treatment of experiments and descriptions of the instrumentation used for hydro- and lithodynamic studies in the nearshore zone.

Some material from these publications, along with results of methodical studies carried out by CMEA countries within the "World Ocean" Program, served as a basis for this chapter, dealing mainly with modern methods of measuring sediment transport in the nearshore zone. The planning of large-scale nearshore experiments, the technical details for implementing instrumentation, detailed mapping techniques, and methods of studying aeolian sediment transport are not discussed in this chapter as appropriate information can be found in the cited monographs.

3.1 Measurement of surface wave parameters

Height, period, and direction are the main surface wave parameters affecting sediment movement. Numerous methods suitable for such measurement of these parameters have been widely analyzed and compared. We confine the discussion to a brief review of the methods based on various physical principles. A detailed specification of various wave meters, together with a vast bibliography on the matter, can be found in Basinski (1989) and NDCP (1988) texts.

3.1.1 Visual observations

Wave height and period are observed visually when instrumental measurements are not possible and approximate values of wave parameters are sufficient. Wave height is generally evaluated with a measuring staff fixed to a pier, a platform, or otherwise. When observing the crest elevation above the trough and the number of waves passing by the measuring point in a certain interval, one can determine the arithmetic mean value of wave height and period. This method is commonly used for evaluation of breaking wave parameters due to the fact that wave meters frequently fail under high dynamic loads and are not easily replaced.

3.1.2 Stereophotogrammetry

This method is based on the analysis of stereophotographs of an agitated sea surface. Pictures are taken with two cameras placed either on the coast, balloons, or helicopters. This method has a certain advantage as it provides a spatially-fixed picture of a wave height distribution across the surface area (i.e., with a balloon raised at 300 m above sea level, this area was 45 m x 270 m (Sasaki, Horikawa, and Hotta, 1976)). This information is very important for studies of storm circulation patterns of the areas under investigation, as instrumental measurements are practically impossible. The expense of helicopter rental and the subsequent processing of stereopictures is the main disadvantage of the method.

3.1.3 Ultrasonic wave meters

These meters act by the echosounder principle. The equipment consists of a pinger and receiver, and is placed either on the sea bottom or above the water surface. Emitted pings are reflected by the sea surface and are registered by the receiver. If the duration of total ping travel, the velocity of sound in water or air, and the depth or height from the calm water level is known, wave height can be determined at the point where the instrument has been placed. Meters of this type are produced in quantity and are widely used for measurements seaward of the plunge point (NDCP, 1988; Luck, 1984). Information is transferred aboard by cable or by radio. In the wave breaking zone, the sea surface is irradiated from the air since bottom pingers cannot be used due to high dynamic loads, signal dispersion caused by abundant air bubbles, and morphological alterations of the bottom surface.

Measurements of surface wave parameters with radar placed in elevated coastal points and sea surface sounding from helicopters (laser methods) are based on the same physical principle.

3.1.4 Devices based on pressure measurements

When surface waves pass above such devices, pressure variations are registered and later converted into sea surface oscillations. Pressure meters of this type are placed on the bottom or in the water column. They are widely used in various countries for nearshore measurements and their modifications differ in the sensitive elements and in the information recording and transferring systems. Testing of this method by Japanese investigators (NDCP, 1988) showed that devices of this type give significant errors when short-period waves are being measured (with period ≤ 3 s) due to signal extinction with depth.

3.1.5 Buoy systems

Buoy systems for wave measurements are based on detection of vertical buoy accelerations along the sea surface, or measuring sea surface oscillations with pressure gauges or ultrasonic sensors placed on underwater buoys.

3.1.6 Wave-recording gauges

Electric wave-recording gauges are based on measuring variations of conductor capacitance or resistance as its moistened length changes with sea surface oscillations. In capacitance wave-gauges, an insulated conductor is placed perpendicular to the calm water surface and variations of capacitance during wave transit are transferred into voltage and recorded. The calibration characteristics of capacitance wave-gauges are usually linear. The conductor is insulated by Teflon or carboniferous polyethylene.

The main element of resistance wave-gauges is a conductor in the form of wire wound about an insulator rod. Variation of the wetted length of the wave-gauge during wave transit results in variation of its resistance.

A contact wave-gauge is a vertical row of contact pairs made of conductive material and placed at a fixed distance (5 to 10 cm) along a rod. The contacts are insulated from each other. As the water level varies, the contacts are either closed or opened and from their known position relative to the calm water level, the surface elevation can be estimated. The accuracy and stability of such wave-gauges are affected by splashes and varying resistance between electrodes because of external impurities.

Electric wave-gauges of these types are widely used for measuring surface wave parameters in the nearshore zone. They are installed on piers, platforms, pile foundations, etc. If such structures are absent, wave-gauges are placed on special tripods or on mobile structures such as sleds, and measurements are taken at a number of points along the underwater slope profile (Coakley et al., 1979; Birkemeier and Mason, 1984; Reimnitz and Ross, 1971). Electric wave-gauges were widely used in the studies of the writers.

3.1.7 Wave direction studies

The physical principles of wave height and period measurement have been considered above. Another important wave parameter is its orientation as the nearshore dynamics depends greatly on the direction of the wave energy distribution.

The direction of wave propagation on a certain portion of the nearshore zone is frequently determined by radar or stereo photography as considered above. Among the in situ measuring devices, electric wave-gauges are the most frequently used: 3 to 5 gauges are placed systematically at points along a certain spatial grid (separated by several meters). Several pressure gauges or buoy systems can also be located along a predetermined grid. Simultaneous recording by all the gauges allows estimation of the intermediate surface slopes from which the direction range can be calculated. Selection of the gauge type and placement depends on the depth, presence of hydraulic structures for mounting instruments, and the measurement precision required for the given task.

It should be noted as a conclusion, that the selection of a method for measuring methods of surface wave parameters in each case depends mainly on task requirements and the availability of technical support at a chosen site. When these are present, electric wave-gauges are used, as they give all the necessary information with good precision.

On coasts devoid of such structures, pressure gauges, stereophotogrammetry, or visual observations can be used. Buoy systems give better results when placed in an outer shore zone. Bottom ultrasonic wave meters should be placed seaward of the wave breaking zone, since the presence of abundant air bubbles in the surf zone causes signal scatter.

3.2 Velocity measurement

Wave flow turbulence in the wave breaking zone, velocity fields caused by wind waves, swell, infragravity and edge waves, nearshore circulation, longshore currents, and compensational water flows perpendicular to the coast are the main nearshore hydrodynamic processes controlling sediment motion and morphodynamics in this zone. These elementary hydrodynamic processes have various temporal and spatial scales of velocity fields, so the methods should provide for a wide frequency range of velocity and direction measurement. The apparatus design should allow for high dynamic loads as well as saturation of the water column by suspended sediment and air bubbles in the plunge zone.

At present, nearshore current velocities and directions are measured at a point (Euler's method) or water flow trajectories are observed (Lagrangian method). For instrument measurements, wave pressure gauges, electromagnetic, and acoustic velocity meters are widely used.

3.2.1 Velocity estimation by wave pressure sensor

Flow velocities can be measured with wave pressure sensors (Vershinsky, 1951, 1952). The measurement is based on recording the hydrodynamic pressure that is proportional to the square of the velocity of the water flow. The pressure-sensing element is a disk (when one component is being measured) or a sphere (when two components are being measured) approximately 2 cm in diameter. The principal aspects of sensor operation, its specifications, measurement errors, and its ability to evaluate turbulence are discussed by Kuznetsov and Speransky (1986) in full detail. Wave pressure sensors provide for adequate measurements of instantaneous values for water flow pulsations at velocities from 0.3 to 1.2 m/s and at frequencies up to 4 Hz. This sensor was widely used during our studies at the Black and Baltic Seas testing sites for measurement of the wave flow velocity field in the wave shoaling and plunging zones. The presence of mobile structural elements resulting in unstable operation in the plunge zone is the main disadvantage of this sensor. WPS sensors can be installed from piers, platforms, or, in their absence, can be placed on rods driven into the bottom. Information is transmitted ashore by a cable.

3.2.2 Electromagnetic gauges

The principle for this measurement is based on Faraday's law, according to which an electric current is induced in a conductor when the surrounding magnetic field is changed. Lenticular, spherical, and cylindrical receiving heads are common. Two electromagnetic inductance coils reside at the center of the head and induce a magnetic field oriented from the center to the periphery. Two pairs of mutually perpendicular electrodes lie in one plane on the head surface. In modern versions of such gauges, current is conveyed to the inductance coil to induce alternating magnetic field around sensor head. When the head flow is around, the adjacent magnetic flux is changed, causing flux change in the inductance coils. As a result, the recorded voltage variations are proportional to water velocities near the sensor head and to the angle between the flow direction and the axis of one of the electrode pairs. In this way, two velocity components are measured. Electromagnetic gauges are widely used and are produced in quantity by a number of companies: Colnbrook (England), Interocean (USA), and NSW (Germany). Prototype gauges of this type have been developed in the USSR without mass production and meteorological testing.

Velocities in the range of ±3 m/s can be measured by mass-produced electromagnetic gauges with an accuracy of 0.5 cm/s. Such gauges are not used for the evaluation of turbulence in wave flow, though they give good results for time-averaged velocities of combined flow and for orbital velocities.

3.2.3 Acoustic gauges

Acoustic gauges are generally used for velocity measurements in shelf and deep water areas (Grant, Williams, and Gross, 1985). In the nearshore zone their use is limited because of their large dimensions (the gauge base usually exceeds 15 cm), the presence of high dynamic loads, and the high bubble concentration in the surf zone. The simultaneous recording of three velocity components at 10 Hz frequency without distortion of the flow structure is a definite advantage of this gauge.

Let us consider the main principles of acoustic velocity gauge operation. Typical instruments have one, two, or three pairs of piezoelectric converters for measuring velocity components. As specific instruments are developed, several methods are used such as the pulse, frequency, and phase difference methods.

The idea of the pulse method is illustrated in Fig. 56. The flow velocity is found from the difference of pulse propagation between two converters. Let the flow velocity be at an angle Q with the axis connecting piezoelectric converters placed at a distance ℓ. If pulses are emitted in turn, the propagation time between converters in both directions t_1 and t_2 is:

$$t_1 = \frac{\ell}{C_0 + V \cos\Theta} \tag{3.1}$$

$$t_2 = \frac{\ell}{C_0 - V \cos\Theta} \tag{3.2}$$

where C_0 is the pulse propagation velocity in sea water. Hence the velocity component parallel to their axis is

$$V \cos\Theta = \ell/2\left(\frac{t_2 - t_1}{t_1 t_2}\right) \tag{3.3}$$

In the frequency method pulses are emitted at a certain frequency, while the frequencies received by the converters and corresponding to the pulse propagation time between them are:

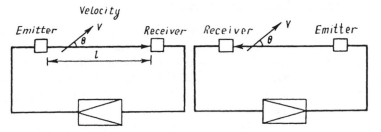

FIGURE 56   The pulse counting method of velocity measurement (NDCP, 1988).

$$\omega_1 = \frac{1}{t_1} = \frac{C_0 + V\cos\Theta}{\ell} \tag{3.4}$$

$$\omega_2 = \frac{1}{t_2} = \frac{C_0 - V\cos\Theta}{\ell} \tag{3.5}$$

Hence

$$V\cos\Theta = \frac{\ell}{2}(\omega_1 - \omega_2) \tag{3.6}$$

Since the ultimate formulas for these two methods do not contain the ultrasonic velocity C_0, they are not sensitive to water temperature variations.

The third method is based on phase difference measurement. If two converters spaced at a distance ℓ are emitting opposing ultrasonic oscillations at a radian frequency ω_0, a phase difference appears because of the Doppler effect and is defined by:

$$\Delta = \omega_0\left(\frac{\ell}{C_0 - V\cos\Theta} - \frac{\ell}{C_0 + V\cos\Theta}\right) = \frac{2\omega_0\ell V\cos\Theta}{C_0^2 - V^2\cos^2\Theta} \tag{3.7}$$

In marine conditions $V \ll C_0$ and thus

$$V\cos\Theta = \frac{C_0^2\Delta}{2\omega_0\ell} \tag{3.8}$$

The main features of acoustic gauges are similar to those of electromagnetic meters. Descriptions of particular versions of acoustic gauges can be found in Podymov and Kos'yan (1982), and Williams (1985).

3.2.4 Lagrangian velocity measurement

Lagrangian measurement of nearshore velocity is used when the pattern and spatial scales of water circulation under storm conditions should be defined for a given section of coast. A solution to this problem with the use of stationary instruments is practically impossible. In this case currents should be studied with the use of various buoys (free-floating, with an underwater parachute in the form of plastic plates) or dye tracers introduced at certain points in the water area. Their temporal migration is observed from balloons, helicopters, or from shore with the help of theodolites or radars. By fixing buoy positions at subsequent time intervals, the velocity along the flow trajectory can be evaluated. The main drawback of free-floating surface buoys is their sailing under strong wind with a greater velocity than that of the actual surface current. Buoys with an underwater parachute do not have this disadvantage, but when entering shallow water, their parachute may be hooked by the bottom. Thus, in shallow water, dyes (such as rodamine or fluorescence) are generally used.

3.3 Measurement of sediment erosion and accumulation at a nearshore site

In the practice of nearshore studies, variations of the sediment layer thickness are commonly estimated by comparing repeated bathymetric maps or profile surveys. A depth survey is usually performed from a vessel with the help of a sinker or echo sounder. The vessel location is determined by the cross-bearings of landmarks on-board, by theodolite intersections from shore, and by other triangulation techniques. On the basis of the survey data, bathymetric maps and plans of the studied area are drawn. Analysis of repeated bathymetric maps and plans or repeated profiles enables one to trace changes of the bottom relief and to locate areas of bottom sediment erosion or accumulation. The time interval between surveys can indicate the intensity of the process.

During storms, profiles can be surveyed from special bridge trestles, tramways, or piers. Such measurements gave interesting data on underwater bar migration and changes in underwater slope relief at different phases of storms of various forces (Aibulatov, 1966; Yurkevich, 1979; Nikolov and Pykhov, 1980; Yurkevich et al., 1982; Nikolov and Kos'yan, 1990; etc.).

More accurate data on relief deformations can be obtained with the use of metallic pins and mobile washers. Divers peg out the special underwater sites with metal pins placed along a chosen route at a given distance (Aibulatov, Kos'yan, and Orviku, 1974; Kos'yan, 1983b). The distance between the top of the pin and the bottom measured during storms shows temporal change at this point. A mobile washer made of sheet metal can be fitted over a pin. When the ground is being eroded around the pin, the washer dips. If erosion is replaced by accumulation, the washer is covered by sand fixing the lowest bottom position in the time interval between measurements.

The method of repeat photography of the same bottom location from a fixed point (Gizejewski et al., 1982) is used more rarely. During storm conditions, observation of bottom sand layer dynamics seems possible only by measurements taken from rigid bridge trestles, piers, and tramways. Using other methods, the data reflect only post-storm bottom variations.

A sensor was developed at the Southern Branch of the Oceanology Institute, which is able to register variations of sediment layer thickness at any point of interest on a reservoir, sea, channel, or river bottom under any hydrodynamic conditions (Phylippov, Kos'yan, and Prokof'ev, 1986; Kos'yan, Onischenko, and Phylippov, 1988). The sensor bites deeply into the bottom and registers variations of pressure on the outer metal diaphragm, caused by variations in the sediment thickness above. Information can be sent ashore by cable or by ultrasonic channel to be recorded, accumulated, and processed.

3.4 Measurement of parameters, direction, and rate of bedform migration

Migration of sand microforms caused by surface waves, wave currents, or channel flows plays an important role in cross-shore, longshore, and channel sediment transport. This component of overall sediment transport has not been studied, as reliable instruments for continuous measurement of bedform geometry are still lacking.

Geometric parameters of bedforms are usually measured by divers (Manual, 1975; Kos'yan, 1983b, 1985b; Miller and Komar, 1980a; Dingler, 1974). The diver descends to the bottom during calm or near calm conditions, and while moving along a set route, measures bedform parameters and estimates their shape and spatial orientation at certain points. Such measurements are usually made with a ruler, compass, depthmeter, and photocamera, which are sometimes united into an instrument called an "underwater pilot."

Observation of bedform dynamics by divers has certain disadvantages: information is interrupted, as measurements during storm conditions are not possible; data on the rate and direction of bedform migration cannot be obtained; and diving operations are laborious and expensive.

Uninterrupted information on the geometry, rate, and direction of migration of bedforms (ripples, ridges, dunes, bars) at the chosen point of the underwater slope can be obtained in the following manner: three or more sensors similar to those described previously are placed by a diver on the bottom in a horizontal plane at the apexes of an equilateral triangle on a metal girder. The construction is put 20 to 40 cm into the ground. The distance between the sensors and their spatial orientation are set beforehand. The height and time of transport for sand bedform crests are found from the time variations of the sediment layer thickness above any of the sensors. The direction, velocity of migration, and length of the bedforms are calculated from the time of crest transit above the sensors.

The aforementioned sensors can be used for routine monitoring of harbors, harbor entrances, and river channel silting, and for continuous observation of the dynamics of bedforms (ripples, ridges, dunes, bars) under any hydrodynamic conditions in zones practically inaccessible for other observation techniques.

3.5 Methods of measuring suspended sediment concentration

There are many methods for measuring concentrations of suspended sediments. The ranges where they may be applied depend on particular conditions: flow depth, accessibility of the observation point, hydrodynamic conditions, degree of sediment concentration, and sediment grain size.

Common methods and commercially-produced devices for observations of suspended sediment in storms are currently lacking. The *Manual on Hydrological Research in the Nearshore Sea Zone...*(1972) does not contain any recommendations on litho-dynamic observations. In the *Manual on Research and Calculation Methods of Sediment Transport...* (1975), a single method is recommended for data collection on suspended sediment motion: sampling by snap-action or long-term filling of sedimentary traps, and even this method is suggested without any explanation of its reliability and suitability for these tasks. Selection of a method or even of an approach for development of new observation techniques calls for a detailed analysis of suspension recording methods and the basic physical principles.

3.5.1 Brief description of available measuring methods

For convenience of discussion, the available methods of field observations should be divided into two groups. The first group should include methods of direct in-flow particle counting or measurement of some parameters related to the variation of particle content in a known volume. The second group should include methods of flow suspension sampling for subsequent analysis. Discussion of the method of the first group is confined to the essential features and evaluation of their fulfillment of the stated requirements.

3.5.1.1 Filming and photography. In accordance with this method, a known volume of water illuminated in place by a "light beam" is photographed. Particles photographed on several horizons are counted. When particle size and density are known, their concentration can be easily found. The main limitation inherent in this method of concentration measurement is that one should know beforehand the law of suspended particle size distribution, which is as difficult to determine as the turbidity distribution. It will be shown below, that the assumption of uniform sediment or constant size distribution with depth is incorrect.

3.5.1.2 Optical methods. Two principal methods for evaluating suspended sediment concentration are based on optical principles: measurement of light attenuation (turbidimetry) and measurement of light energy dissipated by particles at non-zero angles to the incident light direction (nephelometry).

By the Bouguer law, a radiation flux Φ_0 having traveled a distance ℓ in some medium, is attenuated by this medium:

$$\Phi = \Phi_0 \exp(-k_a \ell) \tag{3.9}$$

where k_a serves as an index of light flux attenuation.

Optical instruments are designed to produce electrical signals J_0 and J, proportional to Φ_0 and Φ correspondingly. The relation between the sought-for concentration and measured values of J_0 and J can easily be derived (Onischenko and Kos'yan, 1989):

$$\overline{S} = B \left(\sum_{i=1}^{n} P_i / d_i \right)^{-1} \frac{1}{\ell} \ln(J_0 / J) \tag{3.10}$$

where P_i is the percentage content of fraction with diameter, d_i, n is the number of fractions, ℓ is the base length of the device, and B is the coefficient, which is determined experimentally.

Sources of photoelectric method errors are thoroughly analyzed by Bosman (1984). In a general case, errors of photoelectric measurements can be divided into casual and systematic ones. The sources of casual errors are:

1. The random nature of particle orientation in the illuminated object. The magnitude of this error can be minimized by increasing the averaged time and enlargement of the illuminated object, i.e., instrument base.
2. Instrumental errors resulting from unequal fluctuations of meter readings during J_0 measurement and during recording of the output J signal. Fluctuations of the base length (ℓ) caused by instrument vibration also contribute to the error.

As Bosman's studies showed, selection of turbidimeter design and averaged reading time can reduce the total error to 5%. The sources of systematic errors are:

1. Variability of J_0 parameter in (3.10) during measurements due to electronic device zero drift and water temperature variations.
2. The difference between sediment grain density ρ_s and the density of particles used for device calibration.
3. Deviation of the studied suspension composition from the calibration composition and its variation during observations.

Zero drift of the electronic devices and the effect of water temperature variations can be reduced to a negligible minimum by a proper choice of device design. The degree of ρ_s variation effect depends completely on the stability of mineral composition of the suspension. In a nearshore zone it usually varies insignificantly.

The problem of systematic errors in concentration determinations caused by instability of sediment grain size composition is of extreme importance. The dependence of concentration determination on a continuum limits turbidimeter application in natural conditions. This method has been abandoned because of the assumption that sediment composition variations could be so dynamic and significant that errors in concentration measurement would be of equal or greater magnitude than the value of the concentration itself.

To estimate the effect of sediment composition variations on the measured concentration values, special investigations have been carried out (Onischenko and Kos'yan, 1989). The sediment concentration in the nearshore bottom area ($z = 10$–150 cm) was found to be more dynamic than the suspension composition. Coefficients of variance of concentration were generally 5 to 10 times greater than those of the particle mean size. This holds true within a wide range of observation intervals, from a few seconds to tens of hours. The systematic error of such measurements is about 25% if the suspension composition during phases of storm characteristic (development, stabilization, attenuation) is determined with the help of a turbidimeter at the point of concentration measurement. The accuracy of concentration measurements can be increased by reducing the intervals between sampling for composition analysis and by determining the vertical distribution of the mean sediment coarseness.

The design of the second group of devices, nephelometers, is based on the linear relationship between the intensity of light dJ_1 scattered by a volume dV_1 in the direction forming an angle φ_1 with the incident beam direction, the volume, and its illumination ε (Kozlyaninov, 1961):

$$dJ_1 = (1/4\pi)\beta(\varphi_1)\varepsilon dV_1 \tag{3.11}$$

where $\beta(\varphi_1)$ is the index of scattering in the given direction by a uniform material. If N particles with sizes from r_1 to r_2 are within sight, then

$$\beta = N \int_{r_1}^{r_2} \beta(\varphi_1, r) f(r) dr \tag{3.12}$$

where $\beta(\varphi_1, r)$ is an index of scattering by a particle with the radius r. It can be concluded that nephelometry bears the same drawbacks as does the turbidimetry described above.

A group of optical devices permits determination of both the concentration and sediment grain size by a "shallow angle method." It is known that roughly half of the scattered light flux is concentrated within a narrow angle around the incident monochromatic beam. This angle is inversely proportional to the particle diameter. Thus information on the composition of an irradiated particle group can be obtained from measurements of light flux intensity at various angles. Devices with a shifting detector are based on this principle (Antsyferov and Kos'yan, 1986). With their help, finely dispersed suspensions and small concentrations can be analyzed. Measurements can be made in a few minutes. In principle, this method allows measurement of particle sizes from one to tens of microns (even if laser sources are being used).

Among the specific designs of this group, the optical backscatter sensor (OBS) is of particular interest for nearshore measurements (Downing, Sternberg, and Loster, 1981). It was designed specially for measurements of suspension concentration near the breaking zone. One of the main interferences for optic measurements in this zone is its saturation by air bubbles. The sensor design is based on detection of radiation scattered at angles 110° to 165° from the initial beam direction. The special studies showed that this is the exact angle range in which air bubbles give minimal scatter when illuminated by a light beam. An infrared emitting diode with a 950 nm peak serves as a light source. An infrared emitter was chosen for two reasons. First, since long wave emissions are quickly absorbed in water, distortions caused by vertical interference between adjacent sensors are minimal; and, second, due to rejection of the solar radiation by a special filter, OBS can operate close to the water surface without notable distortions of the desired signal. OBS readings, like those of other optical sensors, depend on the reflectivity of suspended particles and their size distribution. OBS calibration for sediment specific to the site to be studied is therefore obligatory. OBS has a linear characteristic from 0.1 to 100 parts per thousand (ppt) by weight, resolution for sand and silt being 0.1 ppt and 0.025 ppt respectively. Suspension concentration can be measured in approximately 1 cm^3 of water. Due to the small size of an OBS, it can be placed close to the bottom at several vertical points.

3.5.1.3 Radioisotope method. The gamma method, based on the attenuation of gamma radiation by matter contained in an irradiated volume, has recently gained much attention. The measured intensity of a narrow monoenergetic beam J, after passing through a layer of thickness ℓ, is:

$$J = J_* \exp(-\rho_s \mu \ell) \tag{3.13}$$

where J_* is the radiation intensity without passing through matter; μ is the mass coefficient of gamma radiation attenuation, ρ_s is the density of the matter.

Small concentrations of suspension can be defined (Papadopulos and Ziegler, 1965) as :

$$S = \frac{\ln(J_0 / J)}{\ell \rho [\mu_s - \mu_0 (\rho / \rho_s)]} \tag{3.14}$$

where J_0 is the gamma radiation intensity after passing through an equivalent volume of fluid, μ_s and μ_0 are the mass coefficients of gamma radiation attenuation by solid particles and by water, respectively.

The resolution of this method depends on the source strength, and the discrete stochastic nature of radiation, and specifying the minimal exposure time, among them. The safety requirements and design considerations usually cause one to choose low-energetic sources such as cadmium and americium with the exposure time required for reliable measurements being only a few minutes. The sensitivity threshold of the device is rather high, not less than 0.5 g/ℓ, which is one of the reasons for limited use of gamma concentration meters in oceanographic studies.

It can be concluded from the above that radioisotope devices can possibly be used in the most dynamic environment—in a wave breaking zone and sometimes in the near-bottom area of passing waves. Experience in the use of these devices in the nearshore zone, however, is very limited. The most successful gamma concentration meter used by the East European countries is the one designed by a team of Polish and Russian scientists (Basinski, 1980; Basinski, Kasperovich, and Onischenko, 1978).

3.5.1.4 Acoustic method. The effect of absorption of ultrasonic oscillations passing through a liquid containing foreign solid particles is well known. As it follows from the theoretical solution (Rytov, Vladimirskiy, and Galanin, 1938) and from the later models (Urick, 1948; Biot, 1956; Hovem, 1980), the relationship between absorptivity and volume concentration remains linear up to 5 to 8% of impurity content. Vladimirskiy and Galanin (1939) proved this experimentally for particles less than 1 μm, while Hovem (1980) showed it to be true for clay and fine silt particles.

Theoretical and technical developments have appeared recently for concentration meters based on the measurement of ultrasound scatter by sediments suspended in sea water (Vincent, Young, and Swift, 1982; Hey, 1983; Ma et al., 1983; Lynch, 1985; Hess and Bedford, 1985; Varadan, Ma, and Varadan, 1985; Hanes et al., 1988). For this purpose, frequencies from one to several megahertz, i.e., with wavelengths of the order of the particle size, are usually chosen. Available devices (Hanes et al., 1988; Young et al., 1982) work at 3 to 5 MHz frequencies and give the vertical concentration profile at 110 cm from the bottom with good spatial and temporal resolution (1 to 2 cm in the vertical direction).

Considerable ultrasound absorption and scattering by air bubbles prevents wide use of the acoustic method in the surf zone. The wave breaking zone is especially saturated by air bubbles. As the magnitude of the received signal depends not only on concentration but on particle size as well, interpretation errors, like in the above methods, will depend on size distribution in time and space.

3.5.1.5 Conductance meters. By this method, in principle, both concentration and grain size distribution can be determined. It is based on the known effect of the alteration of electrical resistance of a known volume of electrolyte by the passing of solid particles. The method itself and the first devices were developed by Coulter (1953). Most devices based on this principle serve for analysis of samples collected in advance and not in situ.

In devices of this type, the sampled volume of water loaded with suspension is pumped through a calibrated orifice impressed with a constant voltage electric field induced by two plate-electrodes. The appearance of a solid particle acting as an insulator leads to displacement of equivalent water volume and is accompanied by a voltage pulse with an amplitude proportional to the particle volume. Particle quantity and sizes can then be found from the number of pulses and the distribution of their amplitudes. Particle shape and density do not affect measurement results.

As the pulse amplitude to particle volume relationship remains linear only in a certain range of the grain/orifice diameter ratio, devices are equipped with sensor sets. The grain diameter range of the majority of nearshore suspensions can only be measured by several sensors. Hence there is a risk of plugging the orifice with coarser particles and the task of advance sample redistribution in accordance with sensor specifications is difficult. If this limitation could be solved, conductance would open wide possibilities for time-saving and precise analysis.

3.5.1.6 Pulse counting method. Analyzers of this type are based on counting the pulses caused by suspended particle impacts transmitted to a piezoelectric crystal mounted on a steel plate submerged in water (Solov'ev, 1967; Salkield, LeGood, and Soulsby, 1981). The crystal transforms these impacts into pulses proportional to particle mass and velocity. Pulse counting provides for determination of unit sediment discharge at a given horizon. If instant velocity values are measured simultaneously, instant concentrations can be calculated. In this case, the frequency characteristics of the velocity sensor should correspond to those of the particle impact counter. Due to the specific nature of the measurements, this sensor is not suitable for wave flows with alternating velocity directions and is now being used in laboratory studies of steady flows and in the nearshore zone for measurements of the pulsation characteristics of suspended tidal currents (Soulsby, Salkield, and Le Good, 1984; Soulsby et al., 1986; 1987).

3.5.2 In situ traps

Now let us turn to the second group of methods which operate by "trapping." These methods have a certain advantage as along with determination of sediment concentration, they enable one to receive information on the sample grain size and mineral composition and particle settling velocity distribution. However, the main disadvantage of this method lies in identifying unknown sediment conditions.

3.5.2.1 Snap-action traps. With these traps, the mean concentration by volume can be determined with the averaged interval being a fraction of a second. Various modifications of this device were used in natural and laboratory studies. In the nearshore zone they were operated from ashore, being placed on supports, vessels and autonomous stations.

According to Watts (1953,1954) and Petukhova and Ivanov (1965), reliable values of point concentration can be obtained during stable swell averaging over the transit of 1 to 20 waves. Our observations (Pykhov et al., 1982) showed that this interval corresponds to 80 to 100 wave periods, which is considered to be enough for reliable determination of the mean parameters of surface waves and of the mean velocities of wave mass-transport (Speransky and Leont'ev, 1979). This calls for the collection of a great number of samples. It should also be noted that huge devices can distort the near-bottom flow structure. This method is used for concentration profile measurement during certain wave phases or beneath breaking wave crests.

3.5.2.2 Suction-type traps. This type of device includes vacuum traps, siphons, and pump-traps. These are the most common devices used in sea and river research. They are generally used in laboratory experiments, so let us discuss them in greater detail.

Such traps (Antsyferov, 1987; Antsyferov and Kos'yan, 1980a,b) are based on the principle of partial flow suction or diversion into a measuring tank. Solid and liquid phases are then separated. The phase discharge ratio is thus found, which at low inertia of sediment particles corresponds to tank filling time-averaged local concentration. Inertial properties of relatively heavy particles can manifest themselves during the diversion of the hydromixture and its motion along the diversion pipes. Small-volume samplers distort the flow structure insignificantly, and in stationary conditions precise measurements can be taken close to the bottom (at 0.5 cm distance) without local disruption.

Various constructions are used: L-shaped, T-shaped, and straight tubes, disk-shaped samplers and samplers with several orifices along a cylinder side surface, etc. In wave-flumes siphon traps are widely used (Hattori, 1969; Hom-ma and Horikawa, 1963; Antsyferov, Kos'yan, and Longinov, 1973; Antsyferov et al., 1978).

Suction-type traps can be used in field conditions only if extended piers and safe stationary supports are available. Although these traps have definite advantages, are relatively inexpensive, and are easy to operate, there are drawbacks. As mentioned previously, reliable mean statistical characteristics can be obtained only by averaging over about 100 wave periods. Hence, long exposure periods and large tanks should be used. Additionally, to obtain the vertical profile of concentration at 10 or more elevations, the experiment should continue for hours, during which time the hydrodynamic regime may change and give incomparable results.

Another important aspect is the suction rate to orbital velocity ratio at the sampled horizon. Recommendations appropriate for steady flow are worthless in this case. Our methodical experiments showed that the suction rate to overflow velocity ratio is insignificant for disk-shaped samplers and samplers with orifices on the extension side surface (Pykhov et al., 1982).

In many water bodies, sampler protection from plugging by algae, chips, etc., presents a serious problem. In such a case, extensions with suction rates lower than the orbital velocity are preferable as they can be washed automatically by oscillatory water motions. Otherwise, the system should be blown through by a water or air jet.

When using suction-type traps one more condition should be satisfied: the velocity of the sediment-water transit along a pipe connecting the outboard device with a pump should exceed the maximum settling velocity of particles contained in a sample. Otherwise, coarse particles will settle in a pipe and the meter will read low.

As previously mentioned, measurements with suction-type traps were taken from stationary structures which by themselves affect the hydrodynamic regime, at least in their immediate vicinity. Until recently, the correspondence of such observations to the natural, unaffected regime was not discussed, while the special experiments, the results of which will be considered later, have proved this influence to be significant.

3.5.2.3 Suspended sediment traps. Suspended sediment traps are tanks left in a flow for a long period and act like small settling tanks, catching sediment suspended at a given horizon. Such devices were used in the early 1950s as a means of comparing sediment transport intensity at various parts of the near-shore zone (Shapovalov, 1956). Traps were placed along a sea channel (one at the bottom and one in mid-flow) on special stationary posts. After a storm or storm series, the mass of sediment entrapped was registered.

Later, this method was used for detailed study of the vertical distribution of suspended sediment (Fukushima and Mizogushi, 1958, 1959). They used naturally dissected bamboo rods secured on a concrete base. Successful experience promoted development of similar methods in other countries (see, e.g., Basinski and Levandowski, 1974). In most cases they exactly copied the Japanese method, with the only difference being the replacement of a bamboo rod by a detachable metal pipe.

The sample accumulation method has evident advantages—simple, inexpensive, and time-saving—when used in vast water areas. Meanwhile, in the form described in the aforementioned studies, it did not allow estimation of the absolute concentrations. Even its use for evaluation of the relative concentration could not be substantiated, since the methodology of trap performance was not studied in any of the works.

In numerous special studies, various methodical aspects of trap use are validated (Antsyferov, 1987; Antsyferov and Kos'yan, 1980a,b; Antsyferov, Kos'yan, and Onischenko, 1975; Dachev, Kos'yan, and Pykhov, 1980; Antsyferov et al., 1977; 1978; Kos'yan et al., 1982). Some results of these studies are discussed below.

Interpretation of results obtained with suspended sediment traps presents some difficulties arising mainly from the intricate description of suspended flow interaction with a permeable device. As this problem could hardly be solved theoretically, another approach was chosen. Effective designs, convenient to use along with wide and detailed methodical experiments and instrument calibration, were developed to provide operations in marine conditions. Relatively light-weight and inexpensive systems have been constructed allowing for their mobile installation and maintenance both in open water areas devoid of special structures with exposure time being equal to the storm duration, several storms or a whole season, and from stationary posts remotely-controlled during storms. The principal designs are shown in Figs. 57 and 58.

To determine the effectiveness of particle entrapment, numerous laboratory and in situ experiments have been carried out (Antsyferov et al., 1977; Kos'yan and Pakhomov, 1979; Kos'yan et al., 1982) during which the data on material accumulated in traps were compared with the measurement results obtained with the suction type traps and with wave velocities. These experiments showed that the effectiveness of particle entrapment was comparable. Some selectivity in particle entrapment was found—silt particles were entrapped less effectively (by 20–25%) (Voitsekhovich, 1986). Thus, the con-

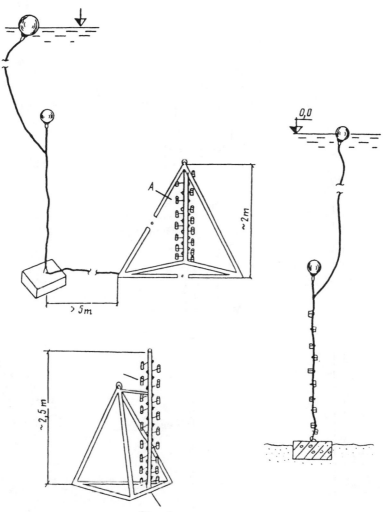

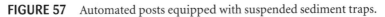

Step - bearing

FIGURE 57   Automated posts equipped with suspended sediment traps.

centration value can be found from the mass of entrapped particles under stationary conditions:

$$\overline{S}_i(z) = \overline{Q}_i(z) / K_0 F_s t_e \overline{U(z)} \qquad \text{for the particles of i-fraction} \qquad (3.15)$$

$$\overline{S}(z) = \overline{Q}(z) / K_0 F_s t_e \overline{U(z)} \qquad \text{for overall particle mass} \qquad (3.16)$$

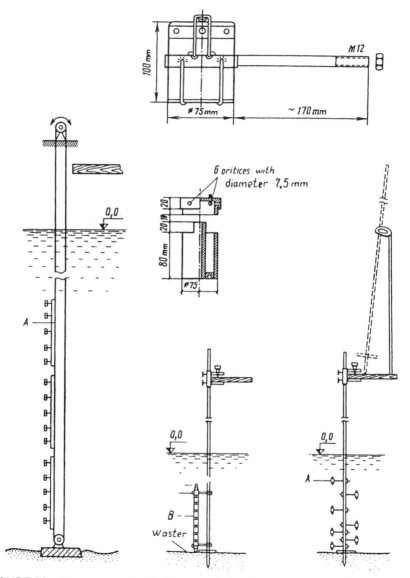

FIGURE 58   The posts equipped with suspended sediment traps serviced during storms.

Here $\bar{S}(z)$ and $\bar{S}_i$ are mean values of the overall concentration and i-fraction concentration at horizon z; $\bar{Q}(z)$ and $\bar{Q}_i(z)$ are the total mass of particles entrapped at horizon z and the mass of i-fraction, respectively; F_s is the projection of the entry orifice square on a plane normal to the wave crest (in accordance with the cassette orientation, the value of F_s may vary within a 5–6% range of the mean value), t_e is the exposure time, $\bar{U}(z)$ is the exposure time-averaged value of the modulus of wave velocity horizontal component at horizon z, K_0 is the coefficient of particle entrapment efficiency with the value of 0.26 (Kos'yan et al., 1982).

Another technique of concentration calculation from the data obtained with traps, more simple and equally accurate, has been suggested by Pykhov and Dachev (1982). It is based on direct comparison of the entrapped particle mass with the concentration measured with suction type traps without accounting for wave velocities. Assuming some deviations to be twofold, approximate estimates can be given:

$$\bar{S}(z) = 0.0148\bar{q}(z) \tag{3.17}$$

or

$$\bar{S}_i(z) = 0.0148\bar{q}_i(z) \tag{3.18}$$

where $\bar{q}(z)$ and $\bar{q}_i(z)$ are the rates of the cassette filling, g/h, by the sediment as a whole and by the i-fraction particles, respectively; $\bar{S}(z)$ and $\bar{S}_i(z)$ are in $g\ell$.

Stability of cassette-type trap performance was tested in a few measurement series taken from a post equipped with four equidistant vertical cassette rows. As a result, the average stability was estimated as 3% for mass and 1.5% for mean size. These values indicate the highest stability among observations taken with various concentration meters.

Justification of trap dimensions, estimates of their interaction, and selection of appropriate spacing were also based on a series of field observations, during which measurements were compared for cassettes placed at various distances in a "herringbone" or "columnar" pattern (see Fig. 58) (Dachev, Kos'yan, and Pykhov, 1980). Observations showed that for the arrangements used, readings of the traps exposed for an hour or more were similar. To exclude their interaction, traps should be spaced at a horizontal distance equal to their height, while in a columnar assembly they can be mounted one upon another without clearance.

To detect the errors related to cassette filling, a series of parallel observations on two vertical rows of traps were carried out. One row was exposed for a few hours, while another one was changed out after one hour (Kos'yan and Pakhomov, 1979). Comparison of the multi-hour accumulation results showed that variation of trap operative capacity does not affect its efficiency until it is filled to enter the orifice level, which is 3/4 of its capacity. The mass of particles in a full cassette should not be considered, while the analysis of sediment composition excluding the upper layer will be sufficiently accurate.

Thus, the suggested traps can be recommended for mass observations aimed at determination of absolute and relative concentrations, and suspended sediment composition on a vast water area during storms.

3.6 Grain-size analysis of bottom and suspended sediment

Microscopy, sieve analysis, electrokinetic, diffusion, filtration, viscosimetry, solution rate, and other methods are widely used now for grain-size analysis of bottom and suspended sediment. Along with sedimentometry, discussed below, these methods characterize sediment dispersion in a wide particle size range.

3.6.1 Microscopy

Particle size can be determined visually or from microphotography if a microscope magnification is known. It can be measured even more precisely with the help of special measuring devices.

To get quantitative data on particle size distribution, particles of different size should be counted within a visible area. Only one dimension (length) is frequently measured. For elongated grains, length and width are measured. With a relatively high number of measured particles (1000 to 2000) a rather accurate distribution pattern could be obtained.

The main disadvantage of this method is that each measurement takes too much time. Microscopy is used mainly as an auxiliary testing method providing for direct observation of sediment particles.

3.6.2 Sieve analysis

The sieve as a means for sorting loose material in fraction size has been known since ancient times. Due to the availability, simplicity, and relative reliability, sieve analysis has preserved its importance up to the present and is widely used in practice.

The sieve analysis principle is simple; a set of sieves with various mesh size is used. In laboratory studies, various sieves are used with mesh diameters varying from a few centimeters to 70 microns. Use of finer sieves seems doubtful. In different countries sieve sets are characterized by various scales, and sometimes even by several scales. In our practice, laboratory grain size analysis was carried out with 18-fraction sieve sets with mesh size ranging from 0.063 to 2.5 mm with a geometric progression index of 1.25 to 1.27.

The precision of sieve analysis depends not only on the number of sieves in a set and their quality, but on other factors, including moisture of sieved material and sieving time. Experience of the user is also reflected in the measuring results.

3.6.3 Other methods

In addition to the principal grain-size studies such as microscope and sieve analysis, other methods are sometimes used, which mostly provide for mean size determination of the smaller sized material, as follows:

1. Filtration and ultrafiltration is similar to sieve analysis in principle, but instead of sieves various porous materials with pore diameters within a certain range are used. A liquid dispersive medium is necessary; vacuum or increased pressure is frequently used to create a pressure difference on both sides of the filters.
2. Nephelometry is based on application of light dispersal laws to solutions and suspensions. Nephelometry is a relative method according to which solution turbidity is found from comparison with a respective standard.
3. The diffusion method is based on application of diffusion laws to colloid solution particles.
4. In the absorption method, the specific surface of a powder can be found from the ultimate absorption of appropriate material by this powder.
5. The method of solution rate is based on a principle according to which the rate of solid body solution is proportional to its surface area.

These methods serve specific purposes and are most frequently used for colloid solution studies. Only absorption and solution rate methods are applicable to coarse dispersed systems.

3.7 Sedimentation method

Sedimentation analysis appeared more than a century ago. Being the most common and available among other types of dispersive analyses, it was used in laboratories at the end of the 19th century and served as the main method of soil and sediment deposition size analysis. It is based on determining the settling rate of dispersed phase particles suspended in a liquid. In the simplest case, particle settling rate is proportional to its size. From the measured rate of particle settling the mean size can be easily found, while the use of appropriate equipment enables one to find a heterogeneous sediment distribution.

The most suitable method for silty sand studies provides for initial concentration and confinement of sediment grains in the upper part of a sedimentation column filled with a liquid, in which sedimentation begins only when a gate is opened.

For practical realization of this method, a special sedimentometer was developed and manufactured in the Southern Branch of the P. P. Shirshov Institute of the Russian Academy of Science. Its description is given by Kos'yan (1990). Methodology experiments with reference samples and natural sands showed that for a given construction, the optimal weight of a sample is 1 or 2 g with the measurement accuracy being within 7–8%.

CHAPTER

4

Characterization of Testing Sites, Description of Experiments, and Observation Data

Due to the variety and intricacy of nearshore dynamic processes, theoretic models and laboratory experiments can serve as a basis for field studies. Validation of laboratory results, applicability of theories, and empirical relationships can be verified only by field observations. Further advances in nearshore sediment dynamics are inseparable from successful natural studies under storm conditions. Therefore, the major long-lasting experiments are an effort to obtain complex observations of intricate and interrelated nearshore dynamic processes. Locations of the study sites are given in the Appendix. Observations in the field, the results of which are discussed in the next chapters, were carried out from 1971 to 1988 in the Black, Baltic, and Mediterranean seas at testing sites located in neutral or slightly accretional areas composed of sands and silts and characterized by a relatively straight shoreline with nearly parallel isobaths. The study regions differ markedly in topography. Observations were carried out during various storm conditions. The common feature of the testing sites is that they all lie in nearshore zones of enclosed tideless seas.

The profiles of underwater slopes and the location of observation points are shown for some of the study areas in Fig. 59. The main characteristics of the observation conditions are presented in Table 5. Additional data on the testing sites can be found in a number of publications (Antsyferov and Kos'yan, 1986; Kos'yan, 1983c, 1984c). All data of field observations of Lubyatowo and Kamchia experiments are kept in archives of the Southern Branch of the P.P. Shirshov Institute of Oceanology, Bulgarian Academy of Sciences (Varna, Bulgaria), and in the Institute of Hydro-Engineering Polish Academy of Sciences (Gdansk, Poland). Data on measurements in the Mediterranean are kept only in the Southern Branch of P.P. Shirshov Institute.

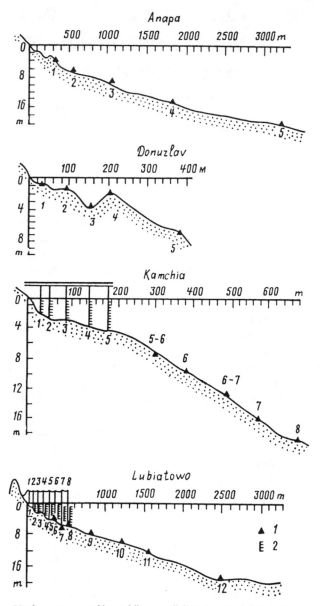

FIGURE 59   Underwater profiles of "Anapa," "Donuzlav," "Kamchia," and "Lubia-towo" testing sites with observation posts. 1: autonomous posts; 2: posts serviced during storms.

TABLE 5   The main parameters of observation conditions on the testing sites

Experiment name and date	Number of storms	Depth range, m	Bottom sediment mean diameter range d_B, mm	Variation range of offshore wave parameters		Processes studied
				H_0, m	T, s	
Anapa-73	4	4.0–20.0	0.10–0.16	1.1–1.6	5.0–6.5	ss
Chernomor-72	1	30	0.06*	2.0	8.0	ss
Donuzlav-75	2	1.1–7.0	0.21–0.36	1.4–1.9	5.5–7.2	ss
Donuzlav-76	2	1.2–4.0	0.31–0.36	0.5–1.0	4.8–6.0	ss
Lubiatowo-74	2	2.3–4.0	0.16–0.46	0.8–1.7	4.3–4.5	ss
Labiatowo-76	4	7.0–18.5	0.14–0.51	0.7–0.8	3.6–5.8	ss,pe,rd
Kamchia-77	3	1.8–25.0	0.15–1.38	0.5–1.2	4.0–6.0	ss,rd,fd
Kamchia-78	1	0.7–4.3	0.20–0.64	0.4–0.9	3.6–6.3	rd
Kamchia-79	2	0.4–2.4	0.42–0.73	0.4–1.5	5.8–7.0	ss
Shkorpilovtzy-82	2	2.0–3.1	0.27–0.54	0.4–0.5	4.5–4.6	fd
Shkorpilovtzy-83	4	2.6–18.0	0.17–0.88	0.6–1.0	3.7–6.0	fd,ss,pe
Shkorpilovtzy-85	3	2.2–17.0	0.17–0.90	0.5–1.3	3.5–6.0	lt,ssrd,fd,tt
Shkorpilovtzy-88	2	0.0–4.5	0.30–1.70	0.6–1.7	4.0–6.5	rd,tt
Sirt-80-81	9	7.0–15.0	0.08–1.45	0.8–1.9	3.2–9.0	ss,fd
Homs-81	2	7.0–16.0	0.21–0.43	0.8–1.5	3.8–6.0	ss,fd
Issyk-Kul-83	0	1.4–7.3	0.18–0.78	–	–	fd

NOTE: ss: suspended sediment; pe: pillar effect on bottom erosion; rd: relief dynamics; fd: bedform dynamics; lt: longshore transport; tt: transversal sediment transport.

*Unsuspended shell present in observation point not included.

4.1 The Black Sea testing sites

4.1.1 Anapa

The study area is located on an accretional form known as Anapskaya spit. The concaved shore has a form of a large-radius arc.

The above-water accretional terrace with sand dunes, wide beach (up to 200 m), and underwater slope with prominent underwater bars are composed by fine-grained quartz sand with minor broken shell. Nearly 90% of sand particles composing the shore and underwater slope are 0.1 to 0.5 mm in size, two-thirds of it being in 0.1 to 0.25 mm fraction. 0.25 to 0.5 mm fraction is predominantly shelly, while 0.1 to 0.25 mm fraction is mainly quartz sand.

The site was suitable for experimental studies for a number of reasons: the thick sand was well-sorted, a considerable portion of the shore was aligned, and the slope had a number of underwater bars typical of shoaling coasts. The investigations were done for irregular wind waves and swell.

4.1.2 Donuzlav

The testing site is situated in Western Crimea, on a shoaling coast composed of shelly sand. The average gradient of the underwater slope is 0.019 (from the water edge to 7 m depth); two prominent longshore bars are present. A series of field observations on concentration distribution and suspended sediment composition was carried out in the bar zone in 1975–1976. Measurements were performed from a 370 m long pier (with supports spaced 12 m apart), extending to a depth of 6 m, and at some distance (30–40 m) from it.

In 1975, observations were made from unmanned posts placed parallel to the pier. Sediment traps were opened at the initial phase of storm development and closed during the storm-damping phase.

In 1976, observations were performed from the pier with the help of rods equipped with sediment traps. The rods were re-equipped at each variation of wave regime.

4.1.3 Kamchia

The Bulgarian Academy of Sciences testing site, Kamchia, was opened in 1977. It occupies the section of the Bulgarian coast with the longest and widest beach. The sand load from the Kamchia River and from minor rivers and streams with a developed gorge system is the main sediment source. Sediment migration is controlled by the wave climate. A north-south longshore storm component is evident.

The topography of the testing site is rather uniform with isobaths nearly parallel to the shoreline. The average bottom gradient from the shore to 18-m depth is 0.023.

A total of 95% of the bottom sediments on the upper part of the slope at water depths of less than 2.5 m is represented by coarse and medium sand. Starting from 12 m depth, the mean sediment diameter grows due to a greater percentage of broken shells. At depths of 15–18 m, like close to shore, more than 90% of bottom particles are coarser than 0.3 mm. The grain size composition of bottom surface sediment is shown in Fig. 60.

In 1977, 1978, 1979, 1982, 1983, 1985, and 1988, comprehensive international research was carried out at the Kamchia testing site, aimed at studying the atmospheric, hydrospheric, and lithospheric interaction in the nearshore zone. This was part of a scientific, collective effort by the former CMEA countries on the project named "The World Ocean." The participants of the experiment were united in international thematic teams. The team responsible for nearshore lithodynamic studies in various years included specialists from the USSR, Poland, Bulgaria, Romania, Cuba, and Vietnam. Research carried out by this team included studies of the vertical distribution of the suspended sediment concentration in the near-bottom layer of nearshore areas varying in dynamic activity, study of the suspended material concentration field in the

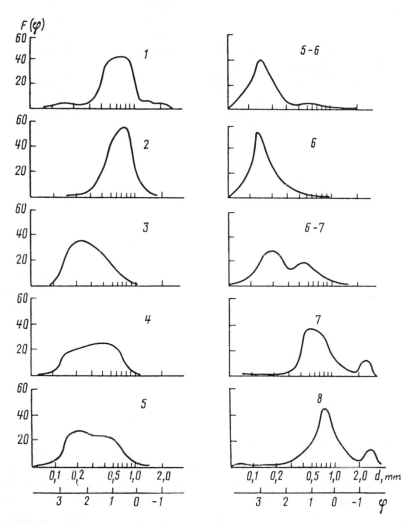

FIGURE 60   Functions of distribution of bottom sediment size at points 1–8 of "Kamchia" testing site.

wave breaking zone and its transformation when crossing this zone, studies of the dynamics of the underwater slope and the bottom sediment composition, study of the dynamics of sand microforms, and improvement of procedures and instruments for the measurement of sediment parameters.

In 1977–1979 the studies were done from a 250 m long pier normal to the shoreline and extending to a depth of 5.5 m. The pier had a cable and telephone communication with the coastal laboratory. Connections were provided for fastening instruments and attaching to the supply line along the pier. For deep-water wave heights up to 4 m, the pier not only overlapped the breaking zone, but intruded into the area of unbroken wave action. Because of the pier, measurements could be taken during storms, with instruments installed at any convenient point along the pier.

Waves were studied from a few points along the pier. Surface wave parameters were also recorded from isolated stationary supports located beyond the pier. These supports were located at 10, 15, and 18 m depths and were connected to shore by a cable.

In February of 1979, the head of the pier was destroyed by a catastrophic storm. Later it was reconstructed and now presents an 80-m long older section with a 20-m long cantilever at its end.

In 1983 a new testing site of the Bulgarian Academy of Sciences Oceanology Institute, Shkorpilovtzy, was opened in the same area. The new testing site includes a modern and safe pier which runs parallel to the old one, some isolated constructions, and a stationary, well-equipped shore station with numerous laboratories. The new pier is 7 m high above the sea level and 200 m long. The depth at its end is 4.5 m. The pier is equipped with all the necessary instruments for fastening sensors, and with reliable radio-telemetric communications, providing direct data input to a computer. Starting from 1983, all the hydro- and lithodynamic studies have been conducted at the new testing site.

The results of the Kamchia and Shkorpilovtzy experiments have been published in three collected books by the Bulgarian Academy of Sciences Publishing House in 1980, 1982, and 1983 under a common title, *Atmosphere, Hydrosphere and Lithosphere Interaction in the Nearshore Zone*, and in a number of articles cited in later chapters.

4.1.4 The underwater laboratory "Chernomor"

The above observations are supplemented by measurements of the vertical distribution of suspended particles obtained from a manned underwater laboratory stationed on the Black Sea floor near the town of Gelendzhik at a depth of 30 m (Aibulatov, Kos'yan, and Orviku, 1974). In the vicinity of the underwater laboratory, a 6.5 m pole was installed in the sand field. Cassette-type sediment traps were distributed over the entire length of the pole spaced 10 to 15 cm apart, and flow meters were placed at several levels.

4.2 The Baltic Sea "Lubiatowo" testing site

The "Lubiatowo-74" and "Lubiatowo-76" experiments were carried out in the Southern Baltic nearshore zone, close to Lubiatowo. Earlier, in 1964–1973, hydro- and lithodynamic processes were studied intensively in this region under the guidance of the Hydraulic Construction Institute of the Polish Academy of Sciences (Gdansk).

The Lubiatowo nearshore zone is characterized by shore-parallel isobaths and equilibrium migrations of longshore sediment flows. Along a bottom profile, four bars typical of the Southern Baltic open nearshore zone lie parallel to the shoreline.

The main feature of the Lubiatowo testing site is a row of eight stationary supports, which traverse the underwater bar zone and reach the depth of approximately 6 m. These supports provide for installation of measuring devices connected to the shore laboratory by aerial cables.

The mean slope of the underwater profile in the bar zone is 0.01, and down to the depth of 18 m it remains constant, being 0.005. The grain size distribution of the bottom sediment is shown in Fig. 61.

Joint projects within the "Lubiatowo-74" program were aimed at studying temporal-spatial variations of nearshore processes and velocity fields, and water and wind energy transformation.

Lithodynamic research during "Lubiatowo-74" was confined to studies of suspended sediment transport. Its ultimate purpose was to obtain data on the distribution of suspended particle concentration above the underwater slope profile in order to test and complement the present knowledge about nearshore conditions during storms.

Studies of suspended sediment dynamics were continued and improved during the "Lubiatowo-76" experiment with the same purpose. In Lubiatowo-76, greater attention was paid to measurements in the shoaling zone seaward of the underwater bars and close to the fourth bar.

4.3 The Mediterranean testing sites

4.3.1 Sirt

In 1980–1982 special studies aimed at developing prediction methods for the suspended sediment distribution in sediment-deficient areas were carried out on a nearshore North African Mediterranean shelf 100 km from Lybian-owned Sirt.

The sediment dynamics of the African Mediterranean shelf is poorly studied, so the results obtained at this testing site, where sediment motion was studied over an intricate underwater topography close to an erosive-accretion shore, are pioneering in many respects.

At this testing site the distribution of the concentration and composition of suspended sediment within the water column and across the area was studied for the first time during storms with sedimentary traps in accordance with the method developed by the authors. Numerous observations of bottom

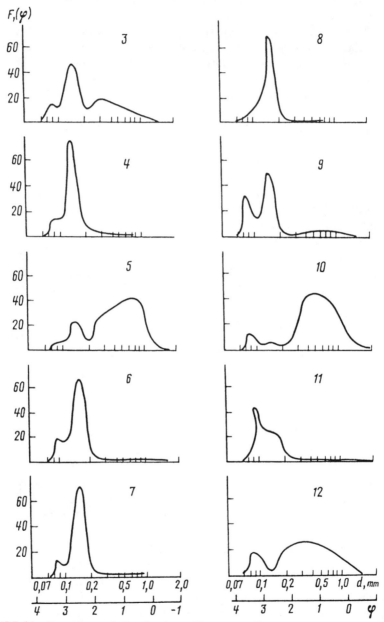

FIGURE 61   Functions of distribution of bottom sediment size in points 1–12 of "Lubiatowo" testing site.

deformation, thickness of mobile layer of sediment of various coarseness, and bed microform parameters were also made.

The studies were carried out on a 4 km section of the shoreline with the 1m isobath as a seaward border. In planview, the shoreline represents alternating bights up to 100 m long and slightly concaved (up to 20 to 80 m). Prominent capes and rivers were absolutely lacking. The shoreline is oriented from WNW to ESE. Nearly 60% of the shoreline is erosive and 40% is accretionary.

Down to the depth of 15 m, the underwater slope is formed by alternating heterogeneous sand fields of varying thickness and of limestone slabs. The sand fields are greatly sloping (approximately 0.010) seaward.

A certain pattern of sediment distribution versus depth can be traced. Between the shoreline and the 8–11 m isobaths, fields composed of silty and fine-grained sands are located. These sediments were formed by the active supply of aeolian material from the African continent. Aeolian sediments are formed by light-brown, well-sorted sand. At greater depths, the bottom is composed of marine sediment formed by abrasion of the bedrock and shell deposits. The grain size of the bottom sediment varies from silty sands to gravels, and is dominated by coarse- and medium-grained, relatively immobile, gray-brown sands. The sharp contrast of these two types of sediment can be traced visually. The specific features of underwater relief and bottom composition of the experimental site are shown in Fig. 62.

Vertical variations of the suspended sediment concentration were measured at 17 points chosen according to bottom features and water depths. The

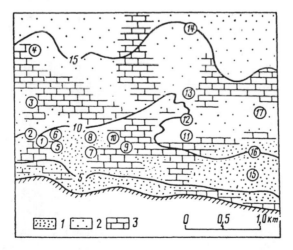

FIGURE 62   Scheme of the experimental site on the North-African coast. Numbers in circles: measuring posts. 1: coarse silts and fine sands; 2: medium and coarse sands; 3: limestone slabs.

locations of these measuring points are shown in Fig 68. Characteristics of the bottom type and installation depths at each point are given in Table 6. The minimum depth for measurement post installation was limited to 7 m, as at lesser depths the tripods were unstable during storms. In most cases, the posts were placed on bottom areas covered with a sand layer more than 0.5 m thick.

4.3.2 Homs

Surface waves are characterized by significant variations of wave elements. The parameters of each sequential wave will be different, and their impact on the bottom and sediment suspension can vary significantly. To give a quantitative evaluation of the concentration profiles of the suspended sediment, one has to operate on effective values of wave parameters.

This aspect was studied during the experimental series carried out in February of 1981 from a petroleum-loading pier on the North African Mediterra-

TABLE 6   Characteristics of the observation conditions on the North African shelf area (see Fig. 62)

| Point | Depth h, m | Type of bed sediment | Ranges of variation of average surface wave parameters during observation time | | |
			H_0, m	H, m	T, s
1	10.1	Rock	1.1–1.6	46–81	5.7–9.0
2	10.1	Medium sands	1.1–1.6	46–81	5.7–9.0
3	11.7	Medium sands	1.1–1.4	46–59	5.7–6.8
4	14	Coarse sands	1.1–1.5	47–71	5.7–7.6
5	7	Medium sands	0.7–1.5	28–64	4.3–8.2
6	9	Medium sands	1.1–1.6	46–77	5.7–9.0
7	8	Fine sands	0.8–1.9	18–64	3.2–8.2
8	9	Medium sands	0.8–1.9	18–64	3.2–8.2
9	7	Rock	0.8–1.9	18–64	3.2–8.2
10	8.5	Rock-Medium sand boundary	0.8–1.9	18–64	3.2–8.2
11	10.5	Fine sands	1.1–1.6	46–81	5.7–9.0
12	11.7	Medium sands	1.1–1.6	46–81	5.7–9.0
13	10.4	Medium sands	1.1–1.6	46–81	5.7–9.0
14	15	Coarse sands	1.1–1.6	47–82	5.7–9.0
15	7	Silty sand	0.9–1.5	39–64	5.4–8.2
16	10	Silty sand	1.1–1.6	46–81	5.7–9.0
17	13.2	Medium sands	1.1–1.6	47–82	5.7–9.0

nean coast close to the Lybian town, Homs. This pier was 1500 m in length, 6 m in width, and extended seawards to 17 m in depth. The height of the pier floor above the sea surface exceeded 10 m, making studies possible in any sea state. The distance between the supports was about 50 m; the work was carried out at equal distances from them to avoid the effects of reflected waves. The bottom sediment beneath the pier consisted of alternating sand beds and rock exposures. Determinations of the vertical distribution of suspended sediment concentrations were made above the thickest sand areas at depths of 7, 10, and 16 m.

Sediment concentrations were determined by the suspension trap technique. From 26 to 52 measurements were made along the vertical line of each exposure. The measurements of the suspended sediment concentration was 6.5 m from the sea floor. The exposure of the traps was determined by the time intervals (3 to 5 hours) when the sea state was considered to be steady, its statistical characteristics remaining unchanged.

4.4 Issyk-Kul

To compare nearshore and lake bedforms, several measurements of the parameters and composition of relic ripples were carried out in the southeastern part of Lake Issyk-Kul in 1983. This part of the lake floor is largely composed of coarse limestone fragments. Therefore, ripples could only be found and measured on the occasional sand-covered patches.

CHAPTER

5

Study of Bed Microforms in the Nearshore Zone

Ripples are the most common bed microforms in the nearshore zone. It has been shown in Chapter 2 that these microforms are generated and exist under a wide range of orbital velocities: from those corresponding to the initiation of sand motion to velocities at which bedforms are eliminated and the bottom is flattened. In natural conditions the complete range is observed during a single storm so it is more difficult to fix equilibrium ripple parameters typical of the given wave regime, than in laboratory experiments. Field studies of bottom microrelief are fewer than laboratory studies as measurements made by divers in the mobile medium under waves are much more difficult. Such works were mainly done in calm weather to study the ripples formed during the abatement of a previous storm. The forms studied were close to the water's edge caused by significantly deformed, repeatedly broken waves, or at relatively deep areas far seaward from the breaking zone.

The scarcity of data obtained with underwater observations for determining the causes of the generation and behavior of bedforms and the lack of observations near the breaking point prompted us to carry out work on different parts of an underwater shore slope over several years. Observations were made, between and during storms, on sites with different intensities of interaction between the water and the sand bottom. Some observations were made under breaking waves, and as far as we know, these observations are unique.

This chapter includes a description of the data obtained and a classification of the results. The available recommendations on the calculation of bedform parameters and the conditions of ripple formation and flattening are verified on the basis of field observations. Some general regularities are clarified in the distribution of bed-sand microforms in the nearshore zones, depending on the nature of the sea waves and bed-sediment composition.

5.1 Characteristics of observed data

Bedform parameters and the composition of bottom sediments were measured mainly at the three experimental sites described in Chapter 4. One of the sites situated in the nearshore zone of the North African Mediterranean shelf was studied in 1980–81. The sandy area of the Bulgarian Black Sea coast was studied during international experiments—Kamchia-78, Shkorpilovtzy-82, Shkorpilovtzy-83, and Shkorpilovtzy-85—of Eastern European countries.

To compare nearshore marine and lake bedforms, relic ripple parameters and composition were studied in 1983 in Lake Issyk-Kul. The observations were made along the underwater profile from the water's edge to the depth of 18 m, with the majority of measurements made within the zone of maximum variability of bed microrelief, at depths of less than 4 m.

When approaching shore, surface waves were transformed through one-fold or repeated breaking. The border between active and relic ripples migrated in depth, depending on wave parameters, and with wave damping, a progressively greater portion of the nearshore bottom became covered by relic forms. The ripples corresponding to wave conditions at the observation moment are regarded as active, while those formed by previous waves are considered relic.

At set times, bed microform presence or absence and their "activity" at the fixed points were observed visually or instrumentally, their parameters were measured, shape and orientation described, and sediment was sampled at various ripple locations.

The ability to reconstruct the shape was adopted as the ripple activity criterion. To evaluate the ripple activity, the seafloor surface was flattened by the diver over an area of 1 sq m at the measurement point. If after 10 min the dimensions of newly formed ripples returned to their values before flattening, these ripples were considered to be active. Note that after a time interval corresponding to the passing of several surface waves, the first reconstructed feature was the length of the sand forms typical of the environment. The process of reconstructing the ripple height took more time.

The major part of the ripple observation data, including surface wave parameters and the characteristics of sediment composition, are given in Kos'yan (1985a,b). Here, therefore, we will consider only the ranges of variation of some measured or calculated values given in Table 7. Some stages of bedform study results have already been published (Kos'yan, 1985a,b; 1986a,b; 1987; 1988a,b,c) and are generalized below.

5.2 Types of bedforms in the nearshore zone

Rolling-grain ripples, though only the relic ones, were observed at depths greater than 12.5 m. It should be noted that such ripples have been observed

TABLE 7   Variation ranges of observed surface wave characteristics, mean diameter of bottom sediment, and bed microform parameters

Name of the experiment	Ripple activity	h m	$\bar{d}$ mm	λ_r cm	H_r cm	$\bar{T}$ s
Kamchia-78	active	0.7–4.3	0.203–0.433	15–300	3.5–40.0	–
	relic	0.7–4.3	0.201–0.642	18–32	3.0–5.5	
Shkorpilovtzy-82	active	2.2–3.1	0.273–0.540	17–45	2.5–9.0	4.5–4.6
Shkorpilovtzy-83	active	2.6–4.5	0.203–0.296	20–250	4.0–50.0	2.7–4.3
	relic	4.5–18.0	0.171–0.877	16–120	1.0–14.0	–
Mediterranean testing sites	relic	7.0–15.0	0.08–1.45	5–130	0.5–29.0	–
	active	7.0–16.0	0.21–0.43	39–90	7.0–23.0	3.8–6.0
Issyk-Kul	relic	1.4–7.3	0.18–0.78	20–40	4.0–9.0	–
Shkorpilovtzy-85	active	2.5–5.0	0.20–0.44	17–45	2.5–10.0	3.5–6.0
	relic	2.5–18.0	0.17–0.90	13–65	3.0–8.5	–

in field conditions for the first time. They appeared as a succession of equidistant sandy bands with flattened crests (the profile of a crest contour appears as a circular arc subtending 90°) and nearly flat troughs (Fig. 63). Their profile shape and low steepness (0.05 to 0.11) nearly correspond to Bagnold's description (Bagnold, 1946). The formation of these unstable ripples can be explained by specific features of the water area. Relatively coarse bed material (mean grain size $\bar{d} = 0.7$–0.9 mm) at these depths was set in motion only by the largest surface waves, while most of the orbital velocities were less than those necessary for initiation of bed sediment motion. Due to this fact, maximal orbital near-bottom velocities did not exceed the limits of the velocity range adequate for the existence of a rolling grain ripple. Under larger waves, ripples with rounded crests are transformed into vortex ripples characterized by a steep profile with wide troughs approximately in the form of a large-radius circular arc, and narrow warped crests (Bagnold, 1946; Carstens, Neilson, and Atlinbilek, 1969; Sleath, 1975; Lofquist, 1978; Allen, 1979; Miller and Komar, 1980a). With each change in orbital motion direction, an eddy formed leeward of the crest rises into water column together with the sediment captured by it. Sand removed from the trough is either redeposited on the adjacent crest or becomes suspended. The majority of ripples observed were of the vortex type. The crests of relic ripples as a rule had a rounded form, while active ripples had a sharper form.

During storms, at considerable distance from the shore and in the deeper water, when nearbed velocities are relatively low, the observed bedforms are symmetrical, two-dimensional sand waves with parallel fronts. Closer to the shore in more shallow water depth, the near-bottom wave velocities become larger. In this case the fronts of the bedforms become curved, with growing

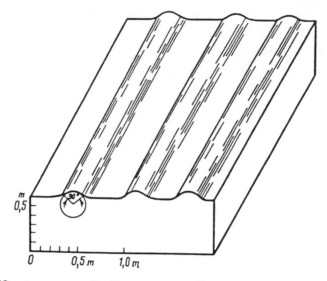

FIGURE 63 Appearance of bedforms corresponding to non-vortex ripple description.

phase shift. Closer to the plunge point, ripples become three-dimensional and cellular without a prominent single front. With further growth of orbital velocities, the bottom flattens. But in the breaking zone itself, specific bedforms develop which differ both in shape and dimensions from the microforms of other sections of the underwater profile. The formation of such forms has not previously been discussed in the literature.

The approximate appearance of breaking zone bedforms is shown in Fig. 64. Crests with a plane surface up to 2 m long alternate with short (not longer than 1 m) and steep (up to 0.4 m) troughs. The ratio of crest to trough length was from 2:1 to 5:1. Their formation seems to be related to hydraulic shock against the bottom due to plunging of the surface wave and the corresponding excavation of bed material. During each half-period of the wave, sediment moves along the propagation path of the wave as a uniform layer several centimeters thick. The sand fraction alternates with vortex strings of rotating particles parallel to each other and normal to the wave front. Plane movement of particles along the crest ceases when the trough is reached, where they are ejected into the water as a cloud or as a funnel-shaped vortex.

Shoreward of the breaking zone, the usual bedforms reappear, first of chaotic cellular configuration, then of more regular shape. At a considerable distance from the breaking point to the shore, the long fronts of bedforms parallel to each other appear again. Ripples become noticeably asymmetric between the water's edge and the line of wave breaking and their orientation changes in such a way that the angle between the water's edge and the normal

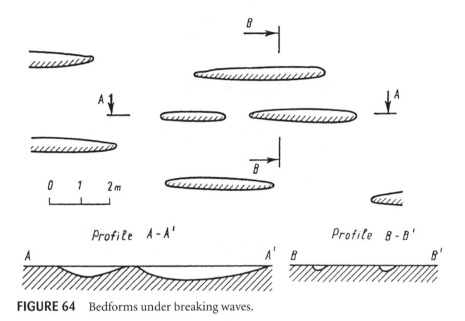

FIGURE 64   Bedforms under breaking waves.

to the sand front becomes more acute than the corresponding angle between the water's edge and the normal to the surface wave front.

The asymmetry of bed forms and the variation of their general propagation direction from the propagation path of surface waves indicate the presence of longshore currents. The degree of asymmetry and deviation show the intensity of these currents. With the help of more detailed observations, a pattern of nearshore currents can be reconstructed if plotted against ripple front and asymmetry orientation. Similar work has been done by Rudowski (1970) for a section of the Polish shore of the Baltic Sea. Observing the nearshore ripples, he managed to locate an area of rip current action.

The distribution of various bedforms and the configuration of their fronts along the profile of the same underwater slope can vary because of storm strength and the location of breaking zones. As the storm abates, sand ripples bear traces of wave impact on the bottom. For this reason, bottom microrelief during a storm cannot be reconstructed from bedform maps plotted in calm weather.

Noteworthy is the peculiar orientation of bedforms in the vicinity of vertical cylindrical piles (about 1 m in diameter). The presence of such piles changes the near-bottom flow direction and increases its turbulence significantly (Antsyferov et al., 1978). In a region of intensive wave action, the near-pile bottom surface (over a distance approximately equal to a pile diameter) was flattened while at greater distance bed forms did occur. Under slight wave action, steeper forms developed near the piles rather than at a distance from them, and their fronts radiated as a fan in all directions.

5.3 Composition of bed microform sediment

It can be noted that the composition of sediment sampled at fixed points on the bottom surface varied significantly throughout the observation period. At the Mediterranean testing site, the mean grain diameter varied within a 50% range, with maximal variations in some points amounting to 570%. At the Kamchia site, mean diameter variation was around 10%, at some points, up to 40%. Other statistical aspects of grain size distribution (standard deviation, asymmetry, kurtosis) varied in even wider ranges. Examples of composition variance during the "Shkorpilovtzy-83" experiment are given in Table 8. In this table, γ_V, K_a, and K_E are the coefficients of variation, asymmetry, and kurtosis, the first, second, and third moments, respectively.

These data show that nearshore sediment mapping based on the results of a single sampling can only be approximate. Along with sediment flow and ripple dynamic studies on various parts of the underwater slope, bottom sediment characteristics should also be determined. Ignoring the time variation of sediment composition (Miller and Komar, 1980a; Tanner, 1971) can result in erroneous interpretation of observational data. During ripple formation, differentiation of sediment composition occurs lengthwise. Different coarse-

TABLE 8   Grain size parameters of sediment sampled at the same points on the sea floor at various dates during the "Shkorpilovtzy-83" experiment

Date	h m	$\bar{d}$ mm	γ_V %	K_a	K_E
29 Oct	2.6	0.27	24.2	−0.67	0.82
6 Nov		0.21	24.6	−0.95	1.59
10 Nov		0.25	21.1	−0.16	0.51
29 Oct	3.0	0.27	28.6	−0.34	0.04
4 Nov		0.24	20.0	−0.43	0.95
6 Nov		0.30	28.0	−0.20	0.51
10 Nov		0.29	26.3	−0.07	0.31
29 Oct	4.5	0.24	27.8	−0.37	0.41
2 Nov		0.25	29.4	−0.50	0.41
10 Nov		0.23	30.0	−0.73	1.33
2 Nov	15.0	0.77	3349	−0.35	0.61
10 Nov		0.82	513.6	−0.05	0.21

ness of crest and trough sediment was noted by Zenkovich (1962); Davies and Wilkinson (1978); and Gillie (1979). Davies and Wilkinson found that finer sediment is accumulating close to ripple crests than in troughs. The difference in sediment median diameter at various points of bed microform can reach 1.5 to 3ϕ, while for 10% of the most coarse particles it can be even greater. This effect was noted only for rather coarse sediment or at relatively great depths of sea water (Gillie, 1979). The composition of fine- and medium-grained sands in ripple crest and trough sediments differed insignificantly (Miller and Komar, 1980a). Some patterns of sediment coarseness variation in ripple troughs and crests can be deduced from our observations.

The relic ripple ratio $\bar{d}_c/\bar{d}_T$ ($\bar{d}_c$ is mean diameter at the crest and $\bar{d}_T$ is at the trough of the ripples) at various depths of the Mediterranean coast is shown in Fig. 65. Values of $\bar{d}_c/\bar{d}_T$ ratios vary from 0.63 to 1.0, most of them being less than a unit and tending to decrease with depth. One should remember that sediment coarseness grows with depth at this testing site.

The above tendencies were also observed for relic ripples and those formed by small waves at the Kamchia testing site: $\bar{d}_c/\bar{d}_T$ ratio decreases with depth and material of relic ripple troughs is generally coarser than that on their

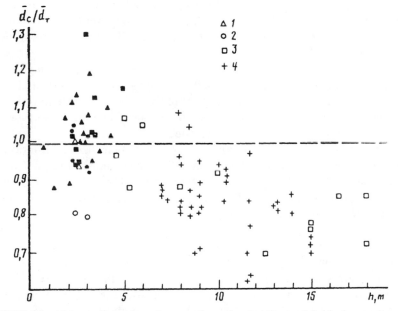

FIGURE 65   Values of d_C/d_T ratio at various depths. Data of field observations: 1: "Kamchia-78"; 2: "Shkorpilovtzy-82"; 3: "Shkorpilovtzy-83"; 4: "The Mediterranean." Dark-colored symbols: active ripples; light-colored symbols and crosses: relic ripples.

crests. The inverse pattern is observed for active ripples: under intensive wave action the mean size of crest sand particles is generally larger than in troughs. Near the breaking zone, the $\bar{d}_c/\bar{d}_T$ ratio reaches especially high values (up to 1.3). The results obtained can be explained in the following way. At the beginning of ripple formation with comparatively low near-bottom velocities, only the smallest particles of a heterogeneous bed sediment are set into motion. Mobile sediment is removed from troughs and redeposited on ripple crests. Less mobile, coarser particles remain in the troughs and at this stage their mean size is greater than that of the crest sediment.

As the waves get larger, an evergrowing part of bottom sediment is mobilized involving material of all the grain sizes. The eddy formed near the ripple trough rises into the water column together with captured sandy particles. Coarser sand removed from the trough is redeposited on the adjacent crest while finer sediment particles become suspended. In this case the crest sediment is coarser than the trough material.

When the storm is over, coarser particles roll from crests into troughs due to gravitation, increasing the mean grain size in the latter. That is why $\bar{d}_c/\bar{d}_T$ ratio of relic ripples is generally less than a unit. The deviation of $\bar{d}_c/\bar{d}_T$ ratio from unity is thus determined by the intensity of wave action on the bottom.

At depths <5 m, in areas of considerable dissipation of wave energy, bottom sediment is usually well sorted (the mean coefficient of variation of the bottom sand is 31.8 ± 1.0%). The grading of crest, slope, and trough material at these depths shows similar values. At greater depths the overall grading of sand deteriorates. In the areas where bottom sediment contains broken shells, the variation of the first moment, γ_{V}, exceeds 100%. Grading of trough material is exceptionally low there. The coefficient of variation of the trough sediment composition can reach values of more than 1000% which considerably exceeds the variation at the crests (Table 9). With a wide range of bottom particle size, $\bar{d}_c/\bar{d}_T$ ratios have their lowest values.

Asymmetry and kurtosis coefficients of the particle size distribution on this area were generally low. Based on the analysis of 145 samples, it was found that $K_a = -0.342 \pm 0.032$; $K_E = 0.466 \pm 0.083$. This means that the grain size composition of the bed sediment may be approximated by a log-normal distribution (Rozhkov et al., 1973).

During storms, the mean grain size of bed sediment increases greatly (by a factor of two or more) near the massive piles (Table 10).

A comparison of bed samples taken at various distances from a pile shows that the coarsest material is deposited near the pile and the mean grain size has its lowest value at a distance of half a meter from the pile. Sediment composition becomes stable at a distance of 3 m and more from the pile.

When an additional source of turbulence of the water mass is present, such as a pile, the number of suspended particles becomes greater (Antsyferov et al., 1978). As finer sand particles become suspended, the coarser particles

TABLE 9   Characteristics of the composition of sediment sampled from different elements of the ripples (by "Shkorpilovtzy-83" experimental data)

Date	h m	Sampled part of rippple	$\bar{d}$ mm	γ_ν %	K_a	K_E
10 Nov	3.5	Crest	0.23	28.4	−0.75	0.99
		Upper slope	0.23	27.3	−0.67	1.15
		Trough	0.20	24.3	−0.58	1.57
2 Nov	5.0	Crest	0.25	26.1	−0.47	0.62
		Upper slope	0.24	25.8	−0.44	0.65
		Lower slope	0.23	25.0	−0.47	0.78
		Trough	0.23	24.9	−0.57	0.83
10 Nov	16.0	Crest	0.82	204.8	0.28	0.34
		Upper slope	0.83	206.2	0.13	0.56
		Lower slope	0.90	392.5	−0.11	0.84
		Trough	0.96	461.6	−0.52	−0.87
2 Nov	18.0	Crest	0.73	125.2	−0.38	0.74
		Upper slope	0.72	106.8	−0.25	1.75
		Lower slope	0.81	202.1	−0.81	1.49
		Trough	1.00	1146.2	−0.42	−0.33

accumulate around the pile. Further from the pile, the turbulence decreases, and suspended particles which saturate the flow settle on the sea floor. Thus, the mean size of sand grains is relatively low here. At a distance exceeding three diameters of the pile, additional turbulence of the water mass becomes insignificant and the presence of the pile does not affect the bed sediment composition.

5.4 Conditions for active ripple existence

It was noted in Chapter 2 that in engineering practice critical velocities are frequently evaluated by a formula such as equation (2.2). It is sometimes used for determination of ripple existence conditions (Murina and Halfin, 1981). The range of ripple existence verified by the data of field observations is graphically shown in Fig. 66. The results of bedform studies in marine near-shore (Inman, 1957; Miller and Komar, 1980a; Nielsen, 1984) and lake zones (Tanner, 1971) were also used for comparison. Inman's observation data (Inman, 1957) are taken from the graphics given in Miller and Komar (1980a) and Nielsen (1981). Only the results obtained in unambiguously deciphered conditions were used.

TABLE 10   Characteristics of sediment composition at various distances from a pile
(results from the "Shkorpilovtzy-83" experiment)

Date	h m	Distance from pile	$\overline{d}$ mm	γ_v %	K_a	K_E
4 Nov	3.5	0	0.66	169.1	−0.22	−0.70
		10	0.31	35.5	−0.52	0.22
6 Nov	2.6	0	0.49	80.2	−0.42	−0.19
		10	0.21	24.6	−0.95	2.59
18 Nov	3.1	0	0.31	28.1	−0.29	0.61
		0.5	0.21	20.7	−0.70	2.22
		3	0.28	20.3	−0.47	1.16
		6	0.28	23.7	−0.28	0.85
		9	0.27	22.8	−0.55	0.41

Tanner (1971) also made marine observations under a calm sea when maximal near-bottom velocities were obviously less than those necessary for initiation of particle movement. In this case only relic ripples were observed; thus, Tanner's data were not used in the present study. Dingler's experimental points corresponding to extremely small ripples (Dingler, 1974) were not used either because of high determination errors. Experimental data from Miller and Komar (1980a) were chosen in accordance with their recommendations. The range of variation measured and calculated parameters taken from observation data are tabulated in Table 11. Amplitude values of near-bottom orbital velocities during ripple observations were calculated in accordance with linear wave theory. In all the comparisons, values of "significant" orbital velocities were used, calculated by heights of the average waves among one-third of the largest waves. In accordance with available estimates (Kos'yan, 1984; Miller and Komar, 1980a; Nielsen, 1981), characteristics of these waves determine the nature of sediment motion under an irregular sea. With the use of these parameters, results of natural observations can be compared with laboratory results obtained under monochromatic waves.

It is evident from Fig. 66 that the data of the field observations do not go beyond the beginning of ripple formation as suggested by Murina and Halfin (1981). U_{cr} curves corresponding to ripple flattening disagree with the observed results. The majority of experimental points suggested by all the authors and proving the existence of ripples under given conditions, lie above the boundary corresponding to the beginning of the "smooth phase" of sediment movement.

The criteria of ripple existence suggested by Brebner (1980) and Dingler (1974) and discussed in Chapter 2, can be written in the form of (2.2). Quartz

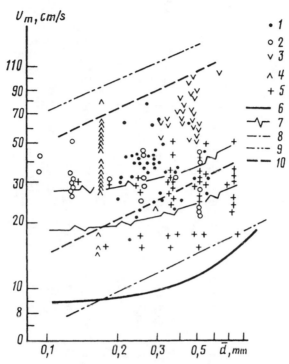

FIGURE 66   Comparison of field observation data of (1) Kos'yan (1985); (2) Inman (1957); (3) Miller (1980a); (4) Nielsen (1984); (5) Tanner (1971) with various relationships restricting the range of ripple existence; (6) initiation of ripple formation, after Murina and Halfin (1981); (7) lower and upper boundaries of initiation of smooth bottom zone formation, after Murina and Halfin (1981); relationships of (8) initiation of ripple formation (Brebner, 1980); (9) ripple disappearance (Dingler, 1974); (10) limiting curves for the range of ripple existence (Vongvisessomjai, 1984).

sands ($\rho_s = 2.65$ g cm^{-3}) have a range of U_m values, delimiting the area of the existence of ripples in the wave flow

$$\text{from} \quad U_m = 2.2\sqrt{gd} \quad \text{to} \quad U_m = 19.9\sqrt{gd} \tag{5.1}$$

Vongvisessomjai (1984) suggested a narrower range:

$$\text{from} \quad U_m = 4.6\sqrt{gd} \quad \text{to} \quad U_m = 14.4\sqrt{gd} \tag{5.2}$$

TABLE 11   Variation ranges of surface wave and bed microform parameters

Condition	h m	$\bar{d}$ mm	$\bar{T}$ s	λ_r cm	H_r cm	Author
Field	0.7–18.0	0.08–1.45	2.7–6.3	5.0–300.0	0.5–50.0	Kos'yan, 1985
	–	0.09–0.50	0.7–13.0	7.8–90.0	0.6–14.8	Inman, 1957
	3.1–21.3	0.17–0.29	6.0–18.2	7.6–27.1	–	Miller, Komar 1980a
	1.3–1.8	0.11–0.61	5.7–12.9	35.0–80.0	4.3–15.0	Nielsen, 1984
	0.2–0.6	0.13–0.71	0.8–6.2	3.5–14.6	0.5–3.4	Tanner, 1971
Laboratory	0.1–0.5	0.20–0.30	1.0–1.5	6.5–24.0	1.5–5.5	Keremetchiev (personal communication)
	0.6–0.7	0.24	1.8–2.4	16.0–18.0	4.0–5.0	Antsyferov et al., 1977
	–	0.28–1.06	1.9–10.1	4.6–33.6	0.1–1.9	Manohar, 1955
	0.3	0.17	3.0–5.0	13.0–16.0	–	Miller, Komar 1980b
	0.4	0.08–0.55	1.0–1.7	2.5–16.6	0.41–2.7	Nielsen, 1979

Having compared the above conditions for the existence of ripples to the data from field studies (Fig. 66), one can easily see that the latter criterion is not fulfilled as a great number of experimental points lie beyond the limits of ripple existence. At the same time, the greater part of natural observation results remain in the range (5.1) which can be used for a rough evaluation of the conditions for ripple formation and flattening.

All the above criteria for the existence of ripples due to wave flow differed from similar criteria for steady flow only in that the maximum value of near-bottom orbital velocity was taken into account instead of steady near-bottom current velocity. It has been shown in Chapter 2 that for correct description of bottom particle behavior, wave period influence should be accounted for (Dingler, 1979; Komar and Miller, 1975; Vongvisessomjai, 1984). Formulas suggested by these authors have been tested; the results are shown in Figs. 67 and 68. Equation (2.33) corresponding to wave conditions at the beginning of

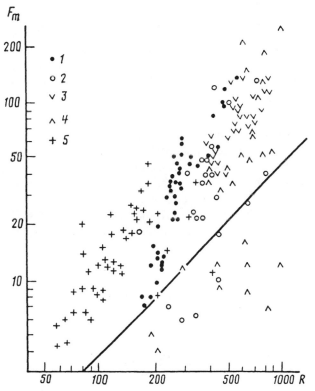

FIGURE 67   Comparison of field observation data with Dingler's relationship (2.27) (Dingler, 1979). For legend, see Fig. 66.

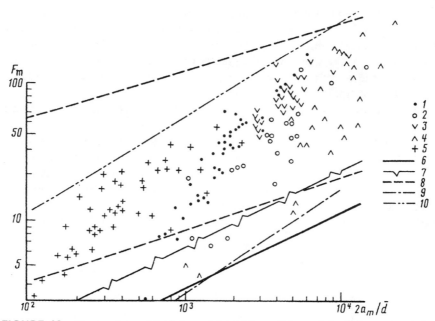

FIGURE 68 Comparison of field data with initial conditions of ripple formation (after Komar and Miller, 1975) (6); initiation of developed ripple (Komar and Miller, 1975) (7); limiting conditions for ripple existence (Vongvisessomjai, 1984) (8); (9) equation (5.31); (10) equation (5.4); (1)–(5) as in Fig. 66.

ripple formation on an erodible bottom, is confirmed by all the observed data. A significant amount of data obtained by various authors do not fit the conditions for the presence of ripples suggested by Vongvisessomjai (1984). Dingler's formula (Dingler, 1979) is the closest to the test points, but it is not supported by some of the observed results.

The set of data on the existence of ripples in the sea (Fig. 68) can be limited in a most efficient manner by the range of conditions found empirically (Kos'yan, 1987).

$$F_m = 3.3 \cdot 10^{-2} \left(\frac{2a_m}{\overline{d}} \right)^{2/3} \tag{5.3}$$

$$F_m = 5.4 \cdot 10^{-1} \left(\frac{2a_m}{\overline{d}} \right)^{2/3} \tag{5.4}$$

It should be noted that for silty sediments the curve given by equation (5.4) approximates Dingler's curve (Dingler, 1979).

The experimental points in Fig. 68, corresponding to the existence of ripple conditions, all lie within a confined area while the above criteria in equations (5.3) and (5.4) border this field only partially. Thus, combining the criteria suggested by different authors and favorably verified by observed data, one can get the range of existence of bottom sand microform.

First, let us compare the chosen criteria with the laboratory experiments published in a number of studies—Antsyferov et al., 1977; Nielsen, 1979; Miller and Komar, 1980b; and Manohar, 1955. The results of this comparison are shown in Fig. 69. Ranges of laboratory parameters are given in Table 11.

Combining conditions (5.4) and (5.5) with the above criteria of Brebner (1980) and Dingler (1974) and allowing for all the experimental points to lie within a range of $2a_m/\bar{d}$ from 100 to 5000, the conditions for ripple existence can be written in F_m and $2a_m/\bar{d}$ coordinates:

$$3 \leq F_m \leq 240 \tag{5.5}$$

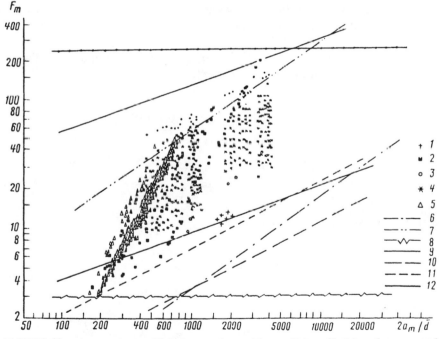

FIGURE 69   Comparison of laboratory data with conditions limiting the range of ripple existence. (1) Antsyferov et al., 1977; (2) Nielsen, 1979; (3) Miller and Komar, 1980b; (4) Manohar, 1955; (5) Keremetchiev (unpublished data); (6) equation (5.3); (7) equation (5.4); (8) Brebner, 1980; (9) Dingler, 1974; (10, 11) Komar and Miller, 1975; (12) Vongvisessomjai, 1984.

$$F_m = 3.3 \cdot 10^{-2} \left(\frac{2a_m}{\overline{d}} \right)^{2/3} \tag{5.6}$$

$$F_m = 5.4 \cdot 10^{-1} \left(\frac{2a_m}{\overline{d}} \right)^{2/3} \tag{5.7}$$

$$100 \le \frac{2a_m}{\overline{d}} \le 25{,}000 \tag{5.8}$$

Conditions for wave ripple existence based on the Shields parameter (Nielsen, 1981; Komar and Miller, 1975) are compared with field observation data with the results presented in Fig. 70, which clearly shows that these authors' criteria are roughly equivalent and mostly agree with the observed data.

Universal conditions of bed microform existence, allowing for the influence of wave period and bottom roughness, can be obtained in Ψ, $2a_m/\overline{d}$ coordinates. All the experimental points obtained in laboratory and natural conditions are shown in Fig. 71.

An attempt to contour the entire field of experimental points by an ellipsis equation (Fig. 71) seems to be more correct:

$$1 = \frac{\left[-5.80 + 0.88\ln(2a_m/\overline{d}) + 0.47\ln\Psi \right]^2}{8.41} + \frac{\left[5 - 0.47\ln(2a_m/\overline{d}) + 0.88\ln\Psi \right]^2}{1.69} \tag{5.9}$$

which can be simplified and reduced to

$$0.23\ln(2a_m/\overline{d}) - 0.26\ln\Psi + 0.06\ln(2a_m/\overline{d})\ln\Psi - \left[0.11\ln(2a_m/\overline{d}) + 0.17\ln\Psi \right]^2 = 1 \tag{5.10}$$

5.5 Active ripple parameters

The results of our measurements of active ripple parameters were compared with those calculated with equations (2.21)–(2.23). Graphical representations

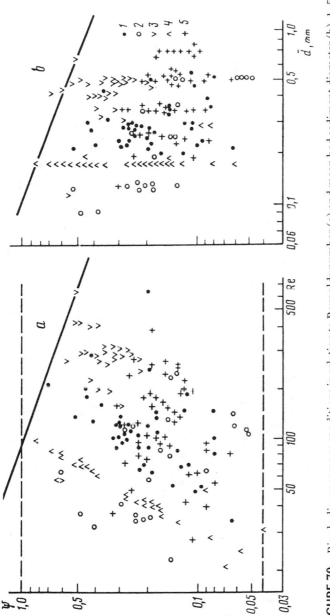
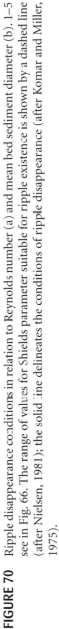

FIGURE 70   Ripple disappearance conditions in relation to Reynolds number (a) and mean bed sediment diameter (b). 1–5 see in Fig. 66. The range of values for Shields parameter suitable for ripple existence is shown by a dashed line (after Nielsen, 1981); the solid line delineates the conditions of ripple disappearance (after Komar and Miller, 1975).

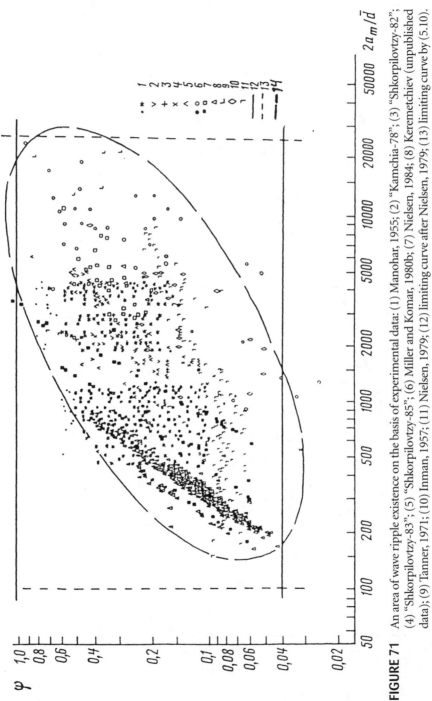

FIGURE 71   An area of wave ripple existence on the basis of experimental data: (1) Manohar, 1955; (2) "Kamchia-78"; (3) "Shkorpilovtzy-82"; (4) "Shkorpilovtzy-83"; (5) "Shkorpilovtzy-85"; (6) Miller and Komar, 1980b; (7) Nielsen, 1984; (8) Keremetchiev (unpublished data); (9) Tanner, 1971; (10) Inman, 1957; (11) Nielsen, 1979; (12) limiting curve after Nielsen, 1979; (13) limiting curve by (5.10).

of such comparisons are shown in Figs. 72, 73, and 74. Fig. 72 suggests that equation (2.17) is indefinite at any constant value. Ripple heights calculated from equation (2.24) systematically exceed the measured values (Fig. 73) (by a factor of 7 on average).

Parameters of active ripples measured under field conditions show the closest fit (Fig. 74) to the curves calculated by Nielsen's formulas by which active ripple length and height can be found from the mobility parameter and orbital diameter calculated from "significant" wave parameters. However, even this degree of agreement between measured and calculated values can be regarded as satisfactory.

5.6 The possibility of wave hindcasting on the basis of paleofacial observations

Recent publications contain numerous attempts to use results from observations of ripple marks in a paleoenvironment to decipher the surface wave parameters in ancient times (Tanner, 1971; Komar, 1974; Allen, 1979; Gizejewski et al., 1980; Allen, 1984; Dupre, 1984). It is assumed in these studies that variations of ripple heights and lengths can be indicative of a near-bottom flow regime, and water depth, wave heights, and periods can be calculated using wave theories. However, all hypothetical calculations only allow

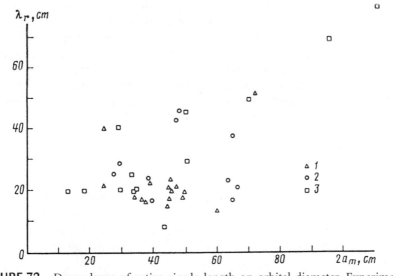

FIGURE 72   Dependence of active ripple length on orbital diameter. Experimental data: (1) "Kamchia-78"; (2) "Shkorpilovtzy-83"; (3) "Shkorpilovtzy-82."

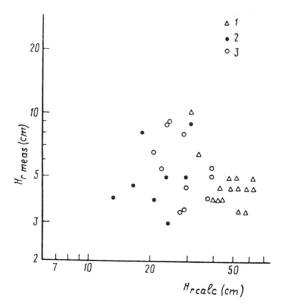

FIGURE 73   Testing of Altunin (1975) formula by field data. (1), (2), (3) as in Fig. 72.

ambiguous interpretations of measurements and they usually differ from experimental data.

The reason for this discrepancy is the fact that as a rule the investigators deal with ripple marks being formed on the sea floor at some moment during damping of a surface agitation, and with the help of these ripple mark parameters they try to determine the conditions of the developed agitation. However, the dimensions of such relic microforms cannot serve as a proper indicator of nearbottom water mass oscillation during severe storms.

Based on field observations, the present study shows that the length and height of relic ripple marks depend primarily on the grain size of the bottom sediments and it reveals the type of special relationships between these values. Field observations were carried out at three testing sites: in the nearshore zone of the North African Mediterranean shelf, on the Bulgarian Black Sea coast, and on the southeastern part of Lake Issyk-Kul. All experimental areas differ markedly from each other by wave conditions. Short-period oscillation dominates in Lake Issyk-Kul, while the sea bed of the African coast is influenced by comparatively long period waves. The periods of the Black Sea waves are generally intermediate. Bottom sediments are sufficiently different for the various sites, but particle density in all cases is approximately the same and averages about 2.65 g cm^{-3}.

It is important to compare the results of these field observations with some results obtained by Tanner (1971) from a water depth of 7 to 19.2 m with sur-

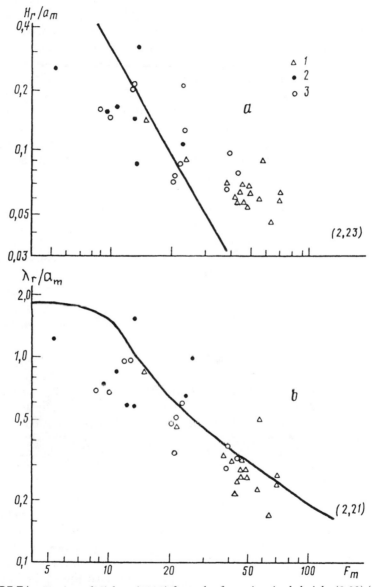

FIGURE 74   Testing of Nielsen (1981) formulas for active ripple height (2.23) (a) and length (2.21) (b) by field observation data. (1), (2), (3) as in Fig. 72.

face wave heights of 24 to 45 cm. In Tanner's works, maximum bottom velocities did not exceed 6 cm/s. Such values are less than the velocity required to
set bed particles in motion. Ripples measured under these conditions can be
viewed as relic ripple marks.

All observations made after storms showed the presence of relic ripple
marks of different size on the sand fields. However, the values of length and
height of the ripple marks were everywhere proportional to the median diameter of the bottom material. The intensity of previous storms had some minor
effect on their dimensions. Parameters of bedforms from the same site
remained typical for those locations during various time intervals. The wave
lengths of the ripple marks vary from 5 cm for silty sand to 130 cm for coarse
sand. Corresponding heights varied from 0.5 to 29 cm, respectively. The relationships between the dimensions of the bedforms and the grain size of the
bottom sediments are practically the same for both the marine and lake environments.

Ripple parameters and mean diameter of the bottom sediment show a
good correlation. The correlation coefficient (r_k) for $\ln \lambda_r$ and $\ln \bar{d}$ value is
0.93 for $n = 203$ measurements, and $\ln H_r$ and $\ln \bar{d}$ gives a value of $r_k = 0.94$
for $n = 174$. With such a high degree of correlation between their coefficients,
$H_r(\bar{d})$ and $\lambda_r(\bar{d})$ can be approximated by:

$$\lambda_r = (83.5 \mp 2.0)\bar{d}^{-1.22 \mp 0.03} \tag{5.11}$$

$$H_r = (17.3 \mp 0.5)\bar{d}^{-1.22 \mp 0.03} \tag{5.12}$$

where λ_r and H_r are measured in centimeters and $\bar{d}$ is measured in millime-
ters. Fig. 75 illustrates the results of the comparison. Accounting for the
dimensional representation of equation (5.11), it can be rewritten in a sim-
pler and more convenient form:

$$\lambda_r = 835\bar{d} \tag{5.13}$$

The results can be explained in the following manner. The length and
height of small bedforms in the nearshore zone depend on the parameters of
the wave field and the characteristics of the sediment particles and the fluid.
During a developing storm they correspond to the wave conditions. At a particular moment as the storm abates, shear stresses near the sea bed become
too small to move separate sand grains and from that time onwards the
dimensions of the ripple marks do not change. This point comes at different
times for sediment with different grain sizes. Critical bed velocities at which
particle movement ceases vary with changing surface wave parameters. How-

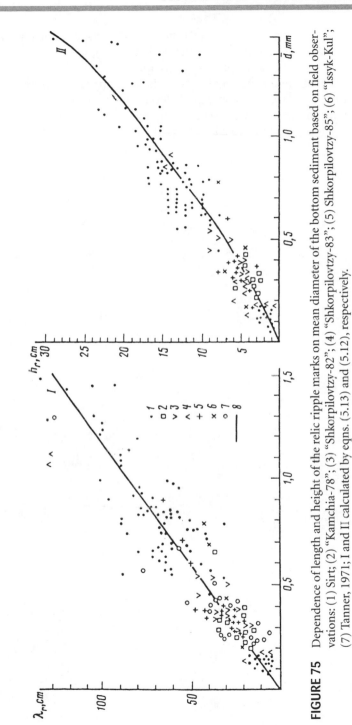

FIGURE 75   Dependence of length and height of the relic ripple marks on mean diameter of the bottom sediment based on field observations: (1) Sirt; (2) "Kamchia-78"; (3) "Shkorpilovtzy-82"; (4) "Shkorpilovtzy-83"; (5) Shkorpilovtzy-85"; (6) "Issyk-Kul'"; (7) Tanner, 1971; I and II calculated by eqns. (5.13) and (5.12), respectively.

ever, these velocities are properties of the bed and they vary about some average values typical of the given area. It follows that relic ripple dimensions for a given ground can correspond only to some minimal bottom shear stress at some point during the damping of the preceding storm. They will never indicate the storm strength at its maximal development. Thus, unambiguous reconstruction of ancient wave-generating environments by relic ripple parameters seems impossible to correctly perform.

CHAPTER

6

Suspended Sediment in the Nearshore Zone

During storms, the majority of the sandy and silty sediment migrates in a suspended state, and this mode of transport has attracted special attention of scientists. When studying the suspension field formation in storm conditions, special attention was paid to:

- Detailed observations of suspended sediment concentration distribution over a vast region including the entire zone of active interaction of waves and sediment;
- Measurements at sites with different physical and geographical conditions and collection of comprehensive data characterizing concentration fields by averaging over various time intervals and taking measurements during various storm conditions;
- Verification and extension of the present-day concepts of suspended sediment dynamics on the basis of the data obtained, and their use for modeling.

Observations of the sea, the results of which are analyzed below, were generally made along a range line, normal to a straight shoreline with isobaths parallel to it.

When studying suspended sediment distribution during a storm on a site with an irregular bathymetry, measuring posts should be located at each characteristic point along the bottom accounting for significant variation of depth and bottom sediment composition (Kos'yan, 1983c). It will be shown later that in the case of a rough bottom relief, sediment concentration is rather non-uniform both along the underwater profile and along isobaths. The volume of suspended sediment depends significantly on the amount of detrital material on the sea floor and on the mean size of bed sediment.

6.1 Distribution of mean concentration values and mean diameter of suspended sediment particles during a storm

Data on the mean concentration of suspended sediments and their composition are primarily needed for engineering applications. Thus, the results of observations of a large region during a few storms enable one to choose the best place to locate a navigation channel, a cooling system for nuclear and thermal power plants, or a waste water discharge. Such observations allow one to estimate the local hydrodynamics and can be used for analysis of the longshore sediment transport, detrital material classification, and sediment composition.

When studying the field of mean suspended sediment concentration and sediment composition, the main goal is finding the solution to the following questions:

1. Up to what depths are sandy and coarse silty particles suspended in significant quantities by storm action?
2. Is it possible to choose any typical area above the underwater profile and to find particular features of the vertical distribution of suspended particle concentration above these areas?
3. What is the role of the breaking zone in the formation of longshore sediment transport, and is it possible to locate a site where most of the sediment flow is occurring?
4. Is it possible to infer any features of the hydrodynamics of the studied water areas using only suspended sediment observations?
5. What are the differences in suspended sediment flow for frontal and oblique shore approach by the waves?

The answers to these questions have been analyzed in detail in a number of publications (Antsyferov and Kos'yan, 1977, 1986; Antsyferov et al., 1977, 1978, 1982; Kos'yan, Dachev, and Pykhov, 1980; Kos'yan, Pykhov, and Phylippov, 1978; Pykhov et al., 1980a,b), so we present only a brief examination of storm-averaged sediment concentration as illustrated in Figs. 76 to 79.

Even a rough analysis of the presented results shows that depths to about 20 m represent a zone of considerable suspension. Experiments carried out from the underwater laboratory "Chernomor" revealed significant quantities of suspension in the nearbottom layer at a depth of 30 m (Fig. 80). It should be stressed that all these samples have a large portion of sandy and coarse silt particles. As all the data were obtained during moderate storms, these depths where suspended particles were observed are obviously far from the ultimate depths.

Some particular features of the vertical profiles of concentration can be noted:

1. In the zone where waves begin to shoal, the vertical distribution of concentration of coarse and medium grained sand are characterized by noticeable

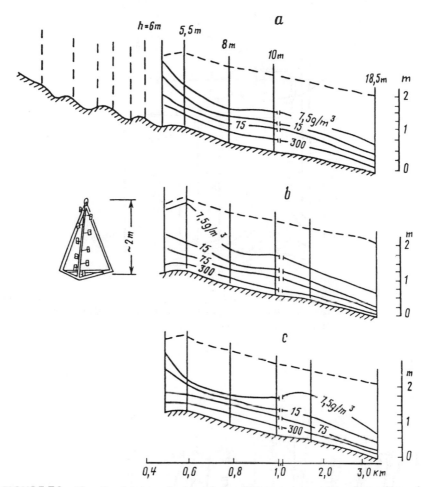

FIGURE 76   The distribution of suspended sediment concentration above the under-water profile of the "Lubyatowo" site during various storms. (1) $\overline{H}_o = 0.7$ m; $\overline{T} = 5.8$ s; (2) $\overline{H}_o = 0.7$ m; $\overline{T} = 3.6$ s; (3) $\overline{H}_o = 0.7$ m; $\overline{T} = 5.1$ s.

variations in the main water column and by a rather significant variation in the near-bottom part of the flow covering tens of centimeters above the bottom. The mean particle size varies considerably with depth in the entire zone.

2. Closer to the shore, as shoaling becomes more pronounced, the distribution of concentration appears to be uniform, or nearly so, in the main water column. It also has a considerable gradient in the near-bottom region. The mean particle size of suspended sediment varies considerably

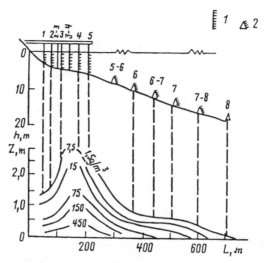

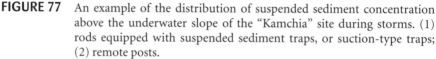

FIGURE 77   An example of the distribution of suspended sediment concentration above the underwater slope of the "Kamchia" site during storms. (1) rods equipped with suspended sediment traps, or suction-type traps; (2) remote posts.

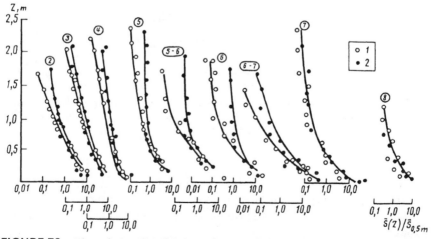

FIGURE 78   The relative distribution of suspended sediment concentration above various sections of the underwater slope of the "Kamchia" site. Location of observation posts as in Figure 77. (1) frontal waves: $\overline{H}_o = 1.1$ m; $\overline{T} = 6.5$ s; (2) oblique waves, $\overline{H}_o = 1.1$ m; $\overline{T} = 5.0$ s.

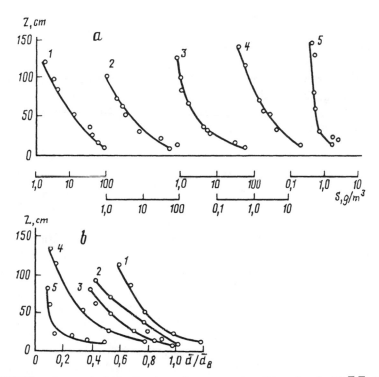

FIGURE 79   The vertical distribution of concentration (1) and ratio (2) $\overline{d}/\overline{d}_B$ for the Anapa site, averaged over a storm period with deep wave parameters $\overline{H}_o$ = 1.6 m; $\overline{T}$ = 6.5 s. The observation posts are located as shown in Fig. 59.

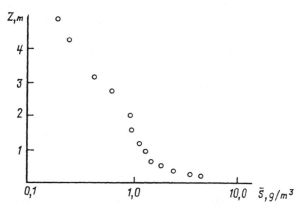

FIGURE 80   The vertical distribution of the suspended sediment concentration within the 5 m near-bottom layer. $\overline{H}$ = 2.0 m; $\overline{T}$ = 8 s, h = 30 m.

in this zone. It should be noted also that due to wash-out of finer fractions from the bottom to the upper flow, mean particle size in the near-bottom suspension can exceed that of the bottom sediment in calm weather (Fig. 79).

3. The features of the vertical distribution of the suspension concentration in the breaking zone are described below. Note that most of the sediment is suspended in the vicinity of the breaking zone for large waves at the developed stages of the storm (point 4, Fig. 77).

Bearing in mind that the velocities of longshore currents also reach their maximum values in the vicinity of the breaking zone, it becomes obvious that it is in this region that the major part of longshore sediment transport occurs. Observation of these maxima is the basis for the determination of the sedimentation rates and the prediction of siltation in marine channels.

A comparison of the general patterns of suspension distribution not only shows the relative characteristics of storms, as can be seen in Fig. 76, but also establishes some hydrodynamic features of the region. During the experiment "Lubiatowo-76," a monotonic decrease of the suspended mass with depth was observed, as expected, along the major part of the profile (not including the breaking zone; see Fig. 76). This pattern is disrupted at depths of 10 to 13 m where all the measurements showed a considerable increase in concentration. This effect can result from significant variations in bed sediment composition or from local variations in hydrodynamic conditions. Where evidence for the first reason is lacking, significant currents should be suggested. It is shown by Kirlis, Massel, and Zeidler (1980) that the mean velocity of the flow oriented along the wave ray reaches its maximum value of 30 cm/s here.

The results of observations during two storm situations with similar parameters of normal and oblique waves (Fig. 78) make it possible to determine the main difference in the nature of suspension for these two conditions. It is determined by the inflow of fine material (coming from the river Kamchia) produced by a more developed flow formed by oblique waves. Owing to this fact, the vertical gradients of concentration throughout the profile are slightly lower during the oblique approach, compared with normally approaching waves, while the mass of suspended particles is higher. These differences are most pronounced in the wave shoaling and breaking zones. Seaward, the variations in the distribution are insignificant, and consists mainly of local sediment. This observation was also supported by the distribution of suspended particle variance, skewness, and Kurtosis (Antsyferov et al., 1980).

It should be added that the influence of pelitic and silty fractions is appreciable only in the upper part of the 2 m thick water layer, and mainly during light agitation. It is obvious that fine particles can be rather active even in calm weather (Aibulatov, Kos'yan, and Antsyferov, 1989).

6.2 Variability of the concentration and mean size of suspended particles during a storm

The variability of suspended sediments during storms is very poorly studied as an aspect of nearshore suspended sediment dynamics. Such data are needed most of all to evaluate the possibility of modeling "instant" suspension concentration fields in the nearshore zone. Knowledge of the variability of suspended sediment concentration and composition is especially important because of migration of the breaking point during storms.

It is obvious that in the vicinity of the breaking zone the conditions for the formation of a suspended sediment concentration field are much more complicated as compared with conditions seaward of this zone. This process can be influenced both by migration of the breaking zone itself and by the fact that this zone acts as a wave filter. Thus, the average parameters for the storm are insufficient for the study of the dynamics of this area. Here, it is necessary to obtain knowledge about the suspension distribution in the studied water areas during various storm phases, and variations of this distribution under changing storm situations.

Let us consider some features of formation of the suspension concentration field in the area of large waves breaking at the developed storm stage (Fig. 81). Here, concentration changes at all depths are directly related to variations in storm intensity. Only one maximum in concentration is observed. This maximum is more pronounced in the near-bottom layer. During the time when the concentration at level $z = 0.1$ m changes by one order, its variation at $z = 1.0$ m constitutes only 10–20% of the previous value. But already at $z = 0.75$ m the variation is of the order of 3 to 3.5 times the previous value. This same tendency is typical throughout the variations in storm intensity. It seems that the water exchange with adjacent areas is more intensive in the upper layers than near the bottom.

It is interesting to compare these trends in the zones seaward and shoreward of the breaking zone. The "Kamchia" experiments helped to obtain information detailed enough for a number of storms to allow such an analysis (Pykhov, Dachev, and Kos'yan, 1980a). A typical example is shown in Fig. 82.

In the course of the storm, breaking of normally incident waves occurred shoreward of point 5. At points 2 and 3, the waves were breaking during the development and the fading of the storm. The area of maximum saturation migrated greatly during the storm. Its position was determined first by variations in storm intensity, and second by breaking zone migration caused by the same variations in storm intensity. In the zone of unbroken waves, the variations in the concentration and size of suspended particles are directly related to variations in storm intensity. At the same time variations in the (d) value are more pronounced here than in the preceding example (Fig. 81) and during other phases of the same storm (points 2 and 3, Fig. 82).

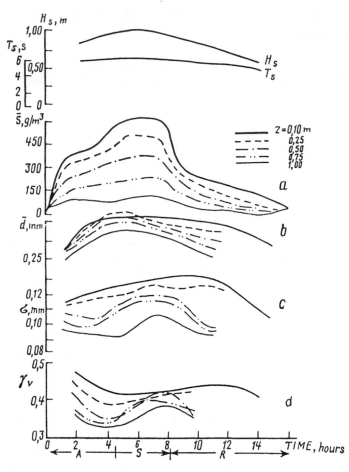

FIGURE 81 Variations of suspended sediment concentration (a), mean size (b), root-mean-square deviation (c), and variation coefficient of sediment composition (d) at various levels during storm on the "Donuzlav" site at a depth of 1.3 m. A,S,R are the phases of development, stabilization, and fading of storm, respectively.

Shoreward of the breaking zone the relationship between suspension processes and storm intensity variations is more complex. Because of the absence of sharp variations in the underwater profile, the area (or areas) of breaking is not permanently fixed at the crests of underwater bars, but shifts actively, following variations of storm intensity. The complex structure of the wind waves leads to a certain "smearing" of the breaking zone in the case of such relief, thus making it rather wide. Then it is more correct to speak about a breaking zone rather than about a single breaking point.

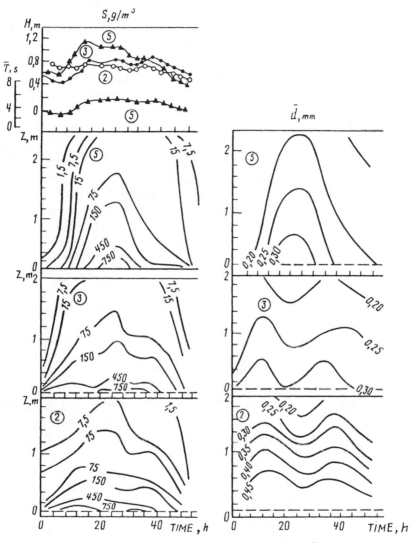

FIGURE 82   Variation of suspended sediment concentration ($\overline{S}$) and its mean diameter ($\overline{D}$) during storm above various sites of the underwater slope of the "Kamchia" area. Figures in circles show the location of observation stations (see Fig. 77).

Point 2 has two pronounced maxima, corresponding to the wave breaking during the storm initiation and fading, and a minimum, coinciding with the phase of the developed storm. During this stage only small waves formed after breaking in more seaward zones reach this point.

Point 3 shows the most interesting case of distribution. The traces left by the retreat of breaking waves at the stage of the developed storm are not very marked here, while those at the fading stage of the storm are rather distinct. At the same time, the waves with medium parameters break here during the developed storm. These are the waves that passed through the filter in the vicinity of point 4 where large waves broke.

It should be noted that fluctuations of $\bar{d}(t)$ values in the breaking zone differ significantly from those in the zone of unbroken wave action. This indicates a different nature of the suspension processes in these zones. It should be stressed that the maximum of the $\bar{d}(t)$ value shows a phase lag relative to that of $S(t)$. The enrichment of the water column by heavy particles, caused by wash-out of light particles, is a process considerably prolonged in time. Since various materials are lifted from the floor and then dispersed at different moments of time, the post-storm composition of surface sediment depends not only on the storm intensity, but also on the duration of storm phases as well, and it varies for the areas seaward and shoreward of the breaking zone boundaries.

The parameters of the two storms, the vertical profiles of the suspended sediment concentration and mean diameter at all the measuring points for two storms of the Kamchia-77 experiment, are given for all the exposures in Figs. 83, 84, and 85. Wind velocity and direction were recorded at the 10 m height on the shore near the trestle. Mean height and period of the waves as well as the corresponding peak of wave energy spectrum were measured at point 7 (see Fig. 59). These measured parameters represent a first approximation to the deep-water characteristic for the two storms. The exposure time of suspended sediment traps (in hours) for the storms 1 and 2 are given in Table 12. For the convenience of comparing the concentration profiles by separate exposures at each point and by each exposure at various points, they have been normalized to the concentration value at the level $z = 0.2$ m.

The following characteristics have been revealed:

1. At all of the measuring points, the vertical gradient of concentration decreases at all levels above the bottom with growth of the average wave height. In the beginning and at the end of a storm, suspended sediment concentration differs threefold in the above bottom layer about 2 m thick, while its maximal gradient is observed in a layer of 0.5 m above bottom. With $\bar{H}$ and $\bar{T}$ growth, concentration gradients decrease throughout the water column and become equal due to greater turbulent mixing of the water measure.

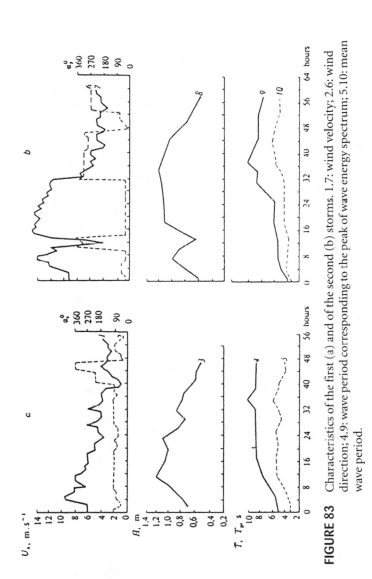

FIGURE 83   Characteristics of the first (a) and of the second (b) storms. 1.7: wind velocity; 2.6: wind direction; 4.9: wave period corresponding to the peak of wave energy spectrum; 5.10: mean wave period.

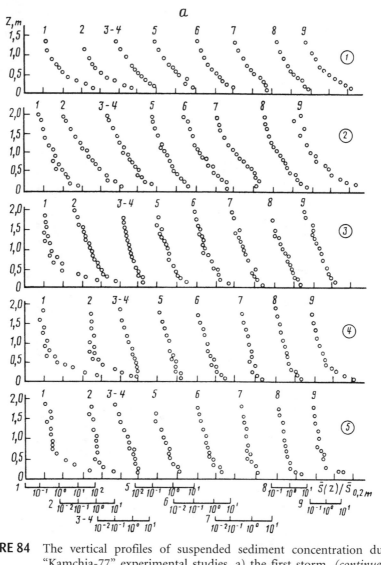

FIGURE 84   The vertical profiles of suspended sediment concentration during "Kamchia-77" experimental studies. a) the first storm. *(continued on next page)*

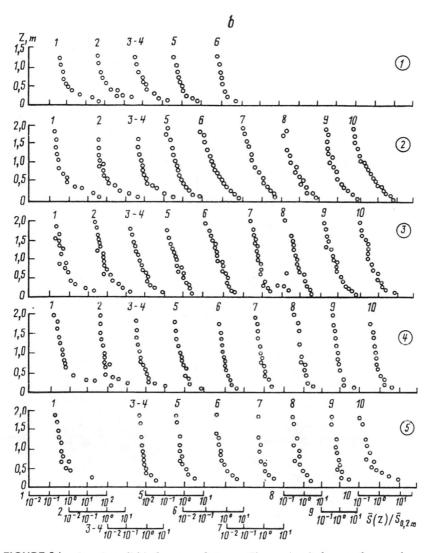

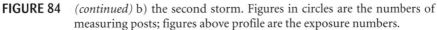

FIGURE 84   *(continued)* b) the second storm. Figures in circles are the numbers of measuring posts; figures above profile are the exposure numbers.

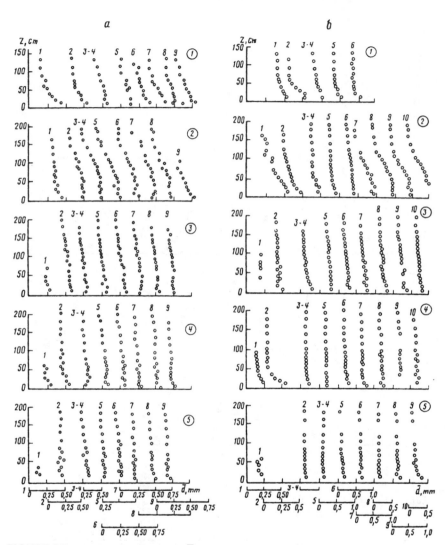

FIGURE 85 Vertical variations of $\bar{d}$ in various points of the underwater slope during storm. For legend see Fig. 84.

2. During the same exposure, suspended sediment concentration gradients increase closer to shore. Had the mean diameter of the bed sediment been constant along the bottom profile, concentration gradients should have decreased at shallower depths due to greater turbulent mixing, just as it occurs at each point with the growth of the average wave height. In our case, the mean diameter of the bed sediment is 2 to 3 times less seaward

TABLE. 12 Exposure time (hours) of suspended sand trap rods

Exposure No.	First storm	Second storm
1	0–5.5	0–5.5
2	5.5–11.0	5.5–11.5
3–4	11.0–23.0	11.5–21.5
5	23.0–28.5	21.5–28.0
6	28.5–34.0	28.0–34.0
7	34.0–39.5	34.0–39.5
8	39.5–46.5	39.5–45.5
9	46.5–52.0	45.5–51.5
10	–	51.5–57.5

than at points 2 and 1. Greater turbulent mixing on a regular basis results in greater mean size of the suspended sediment at lesser depths and, consequently, in greater concentration gradients in the upper layers.

3. The curves of the vertical distribution of the suspended sediment mean diameter are similar at all the measuring points despite the various absolute values of $\bar{d}$. The near-bottom $\bar{d}$ gradient is somewhat higher in the beginning and at the end of the storm only at points 1 and 2. At all of the measuring points and in all the exposures, $\bar{d}$ values generally decrease with z height: two times at points 1 and 2, and by 10–20% at other points within a 2 m thick layer.

6.3 Low frequency variations of suspended sand concentration and composition

The significance of low frequency waves on sediment transport and nearshore zone morphodynamics has been noted in Chapter 1. Low-frequency oscillations observed during wave height and velocity measurements have not been detected during the concentration measurements for a long time, as the instruments and methods used could not provide for prolonged records with discretization suitable to separate such oscillations. The optical backscatter sensors (OBS) and acoustic concentration meters developed in the early 1980s made such recordings possible. The very first measurements taken with OBS in natural conditions permitted separation of the elements of low-frequency modulation of the suspension concentration (Downing, 1984; Sternberg, 1984). More recently the C^2S^2 experiments (Hanes and Huntley, 1986; Hanes et al., 1988) on the Oregon Coast of the USA (Beach and Sternberg, 1988), and our observations of the Kamchia testing site (Onischenko and

Pykhov, 1990) have clarified some features of the low-frequency variations of concentration.

Hanes and Huntley (1986) measured the suspension concentration at 5 levels, from 2 to 18 cm above the bottom at a depth of 1.1 m, which is seaward of the wave breaking zone. The maximum wave height was 0.25 m with a period of 6 to 8 s. Under these conditions, the sediment was suspended only near the bottom (2 to 6 cm) while at $z = 13$ cm, the sediment concentration was approximately zero. The intensity and height of suspension grew rapidly during the passage of a wave train with a 2 to 3 min period. The concentration value during these periods mainly determined the value averaged over 15 to 20 min. Since the data on suspension composition is lacking, the source of the suspension cannot be determined, whether it was formed through local erosion or was brought from outside the measuring site.

The results of simultaneous measurements of two horizontal components of velocity at $z = 31$ cm, wave heights, and suspension concentration at 4 levels (from 5 to 26 cm) are shown in Fig. 86 (Beach and Sternberg, 1988). The low-frequency variations in these fields are clearly seen. The significant wave height during this interval was $H_s = 0.77$ m and $H_s/h = 0.57$. It can be clearly seen that during the first 5.5 min strong water movements (up to 2.5 m/s) with a period of about 2 min are dominant, while in the second half of the recording, wave motions with a period of 8 s prevail. The peaks in suspension

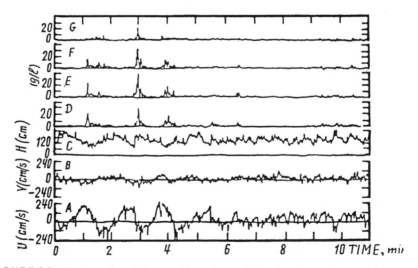

FIGURE 86   Temporal variations of longshore (a) and normal (b) components of velocity at the level of 31 cm above the bottom, wave height (c) and suspension concentration at $z = 5$ cm (d), $z = 14.0$ cm (e), $z = 26.0$ cm (f) (Beach and Sternberg, 1988).

concentration each lasting about 30 s, amounting to 26 g/l with a period of approximately 2 min, also occur in the first 5.5 min. The comparison of concentration profiles showed that during the first half of the recording, when infragravity wave motion was observed, concentration values were 5 times greater than during the second half, devoid of such motion.

Unfortunately, the principle of measuring backscattered emission by suspended particles used in the above studies, does not provide for evaluation of variations in the suspension composition. To estimate the possible low-frequency variations of suspended sediment concentration and composition in field conditions, we used suction-type traps (Pykhov et al., 1982) with sampling discretized to 12 to 15 s at a measuring period of 30 to 35 min or to 2 min for a measuring period exceeding an hour. All the results given below were obtained in the breaking zone at $z = 0.1$ to 0.3 m above the bottom.

An example of a temporal variation of concentration and suspended particle mean size for swell with $\Delta t = 2$ min is shown in Fig. 87. It is seen that the period of concentration fluctuations is 8 to 10 min. Deviation of the concentration around the series-averaged value of 2.9 g/l does not exceed 2.0 g/l. Similar concentration variations with an 8 to 10 min period were observed during other recordings, both for swell and wind waves. The physical nature of this periodicity is hard to explain since the data on wave velocities measured in the corresponding temporal range are not presently available.

Typical examples of temporal series for the suspended sediment concentration and mean size with $\Delta \tau = 13–15$ s and duration of 15 min are shown in Fig. 88, and with a duration of 35 min in Fig. 89. These data show that in the wave breaking zone, intensive sediment suspension under wind waves has the form of sharp concentration peaks 20 to 60 s long with concentrations 3 to 5

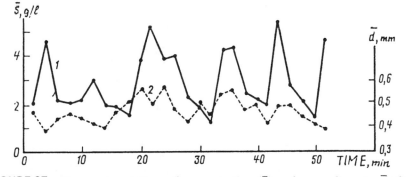

FIGURE 87   Temporal variations of concentration ($\bar{S}$) and mean diameter $\bar{d}$ of the suspended sand in the wave breaking zone. The measuring level is 15 cm above the bottom; the water depth is 1.2 m. Sampling is discretized to 2 min. 1: concentration in g/l; 2: mean diameter in mm.

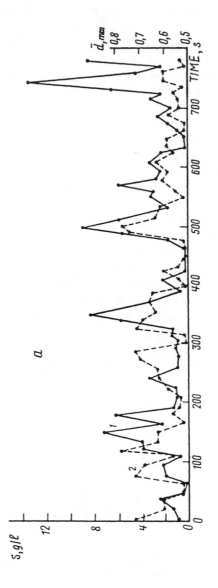

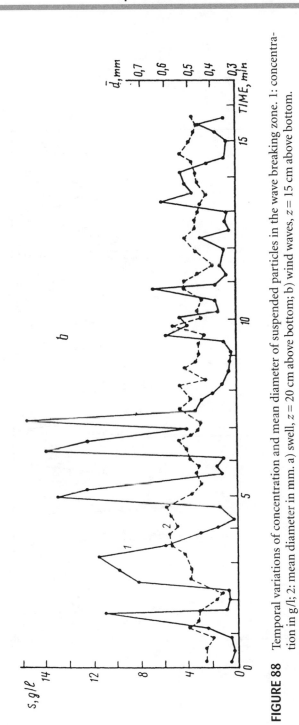

FIGURE 88   Temporal variations of concentration and mean diameter of suspended particles in the wave breaking zone. 1: concentration in g/l; 2: mean diameter in mm. a) swell, $z = 20$ cm above bottom; b) wind waves, $z = 15$ cm above bottom.

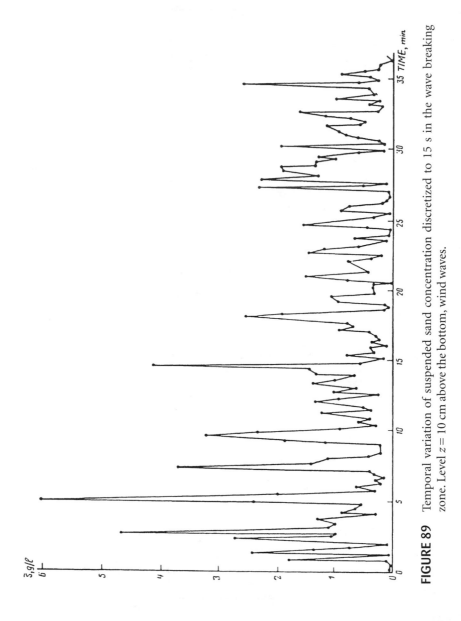

FIGURE 89   Temporal variation of suspended sand concentration discretized to 15 s in the wave breaking zone. Level $z = 10$ cm above the bottom, wind waves.

times greater than the recorded time-averaged values for the series. Thus, in Fig. 87 the total duration of the peaks is only 30% of the complete series, while their share in the recording-averaged concentration amounts to 60%. This means that these particular peaks mainly determine the average concentrations during 10 to 15 min intervals. The periodicity of concentration peaks is 2 to 3 min. This value agrees with other measurements (Beach and Sternberg, 1988; Hanes and Huntley, 1986; Hanes et al., 1988) in which the periodicity of concentration peaks were found from the infragravity oscillations of the velocity field.

In contrast to concentration, the suspended particle mean diameter varied less than 25% about the recording-averaged value. It should be noted here that in roughly half of the cases (as, for example, is shown in Fig. 88a), the suspension concentration correlates well with the particle mean diameter, while in the other half (as in Fig. 88b) such correlation is absent. A rigorous physical explanation for this behavior would become possible only if the recordings of the hydrodynamic characteristics at the measuring point were equally elaborate. Without them, we can assume only that in the series shown in Fig. 88a, sediment particles suspended from the bottom close to the measuring point predominate. In this case the greater hydrodynamic wave effect will result in a simultaneous increase in the concentration and mean size of the suspension. This assumption is confirmed by the suspension composition histograms (Fig. 90a,b) corresponding to minimal and maximal concentration values. It can be easily seen that the greater mean size of suspension particles corresponding to the concentration peaks is caused by the local erosion of fractions coarser than 1 mm which are not in suspension at lower concentrations.

In the series shown in Fig. 88b, the mean particle size grows with concentration only in the first 250 s of the measuring interval. During this period, histograms for the extreme points of concentration (Fig. 90c,d) are similar to those in the above case. At $t > 250$ s, correlation between the suspension concentration and the mean size is not observed. Histograms for the suspension particles of extreme concentration values after 250 s are practically identical (Fig. 90e,f). Dominance of the suspension brought from the adjacent sites over the local erosion is a possible explanation. At this site, the waves approach obliquely and the wave height varies visually along the shore. This combined with the cuspate, jaggedness of the shoreline, suggests nearshore circulation patterns which may be responsible for the observed variations. The probable influence of circulation on the concentration is shown in chronogram form in Fig. 91. This concentration was measured with an averaging internal period of $\Delta t = 2$ min at the wave breaking point with almost normal wave approach. The depth at the measuring point was 1.2 m. The sampler was placed at 10 cm distance from the bottom. The mean bed sediment diameter was 0.35 mm. Throughout the observation period, the mean diameter of the suspended sediment varied insignificantly relative to its value averaged for 1.5

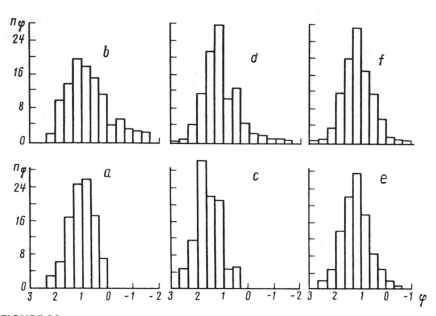

FIGURE 90 The histograms of suspension samples at the minimum and maximum concentration (a,b) for Fig. 87a; (c,d) for Fig. 87b, $t < 8$ min; (e,f) for Fig. 87b, $t > 8$ min.

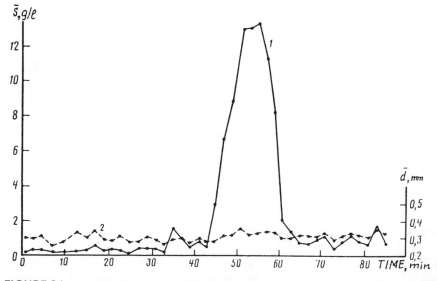

FIGURE 91 Temporal variations of concentration and mean diameter of suspended particles in the wave breaking zone, $z = 10$ cm above bottom, depth of 1.2 m, sampling discretized to 2 min.

hours. At the same time, a unique peak of abnormally high concentration (of an order of magnitude) of 20 min duration is observed on the concentration chronogram. Visual observations suggest that this concentration peak corresponds in time to the rip current action at the measuring point. This current was marked by strongly turbid water. During measurements, the shoreline was jagged with cusps, and circulation cells were clearly seen.

It should be noted as a conclusion that the above results are only an attempt to clarify the temporal variations of concentration in the wave breaking zone. The authors realize that these results are mostly qualitative, as the duration and discretization of sampling do not permit statistical analysis. Nevertheless, these data suggest some insight into the low-frequency variations of the suspension concentration and composition, practically unknown until recently.

The future research should be aimed at recording of prolonged series (about 2 hours) with the shortest possible averaging time for separate samples. This will make a detailed quantitative analysis possible and will permit plotting the suspension concentration and composition ranges. To develop physical models of the temporal and spatial variations of the suspension field, measurements of the characteristics of velocity and suspension fields should be made simultaneously at several levels.

6.4 Distribution features in the breaking zone

Different types of vertical profiles of concentration exist for the following wave regimes: weakly deformed waves, greatly deformed but unbroken waves, spilling breakers, plunging breakers, and the bore zone. The properties of these profiles can be seen in Fig. 92.

In the zone of weakly deformed waves most of the suspended sediment is in the near-bottom layer. As wave height increases, suspended sediment is distributed over the whole depth with a large gradient in the near-bottom layer.

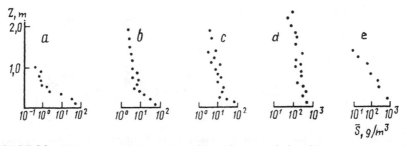

FIGURE 92   The types of vertical profiles of suspended sediment concentration under: a) weakly deformed waves; b) greatly deformed but unbroken waves; c) spilling waves; d) plunging waves; e) in the bore zone.

In the zone where waves break by spilling, the distribution of sediment is almost uniform with a slightly increasing gradient in the near-bottom layer. In the zone of plunging breaking waves, the distribution is almost uniform over the entire depth. In the bore zone the sediment is distributed with a constant gradient in semi-logarithmic coordinates.

It should be noted that the breaking zone is also distinguished by the characteristics of suspended sediment composition. The vertical distribution of the main composition parameters is almost uniform while at the adjacent sites these parameters vary significantly (Fig. 93). It can be assumed that the post-storm bed sediment composition in this zone will be more uniform than outside the zone.

In Fig. 94, examples of the concentration profile during a storm are given at the same point of the underwater slope while the seaward border of the breaking zone migrated relative to the measuring point (Kos'yan, Dachev, and Pykhov, 1980). The first profile was obtained when the weakly deformed waves were acting upon the bottom. The profiles 2, 3, and 9 were obtained when the greatly deformed unbroken waves were passing through the measuring point. For profiles 4 and 8 the point of observation was situated at the beginning of the breaking zone. For profiles 5 and 7 the point of observation was in the

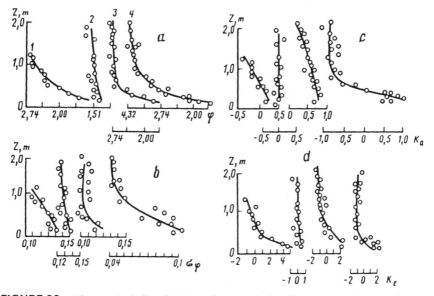

FIGURE 93   The vertical distribution of suspended sediment parameters. a) mean diameter in φ units; b) root-mean-square deviation; c) skewness; d) kurtosis. 1: bore zone; 2: plunge point; 3: greatly deformed waves; 4: weakly deformed waves.

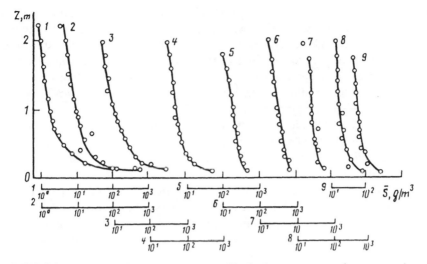

FIGURE 94   Changes of concentration profile during a storm at the same point of underwater shore slope. 1: weakly deformed waves; 2,3,9: greatly deformed but unbroken waves; 4,8: the point at the beginning of the breaking zone; 5,7: the point at the middle of the breaking zone; 6: the point at the end of the breaking zone.

middle of the breaking zone. For profile 6 the point was at the end of the breaking zone. In Fig. 93, it is seen that if the degree of wave deformation is increasing, the suspended sediment concentration gradient is decreasing.

During the experiment "Kamchia-79," an observation series was carried out to determine whether the variation of sediment concentration and composition in the zones of intensive wave action is an ergodic process; and if it is, what time is needed to attain the stable mean parameters (Kos'yan and Kochergin, 1987). Water with suspended sand was sampled continuously with a suction-type trap at 10 cm above bottom. Under the calm sea the depth of the observation point was 2.4 m. The choice of the observation site provided for the maximal variety of conditions. During the experiment, the majority of waves (65%) broke 3 to 5 m shoreward of the sampling point, while 25% of waves broke directly at this point. Ten waves out of a hundred approaching the observation point had already broken. The maximal wave height in deep water amounted to 2 m with the mean period being 6 s. The experiment was carried out during the stable storm conditions and it lasted 106 min. During this period the suspended sediment was sampled continuously. Each sampling took 100 to 140 s, 2 min average. During this interval, 30 to 60 l of liquid entered the trap.

During the experiment the concentration of the suspended sediment varied from 0.17 to 2.55 g/l with the mean concentration for the whole observation period being 0.72 g/l. Faired curves of temporal variations of these values are given in Fig. 95. Sequential averaging in time shows (Fig. 95) that the mean concentration becomes practically independent of averaging time after 80 min, which roughly corresponds to the passage of 800 waves. Deviation of the concentration values averaged over a longer period is less than 15%. The stable mean values of the suspension composition parameters are attained very quickly; when averaged over a period of more than 50 min, their fluctuations are less than 1%. Fig. 96 illustrates the dependence of mean parameters shown in Fig. 64 on averaging time. The large measuring time needed to obtain the reliable average values of sand concentration is explained by low-frequency fluctuations with a period of 10 to 15 min (Fig. 94a).

The breaker zone is very important to the distribution of suspended and bed load sediments throughout the coastal zone. Studies by Kos'yan (1985a,b) give a detailed survey of the concentration, distribution, and the composition of sediment suspended in the vicinity of the plunging area and in the bore

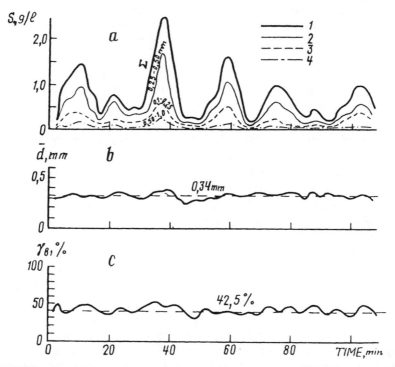

FIGURE 95   Temporal variations of the suspended sediment parameters. a) overall sand concentration by fraction; b) mean diameter; c) variation coefficient of suspended sediment composition; 1: overall concentration; concentration by fractions: 0.25–0.5 mm; 3: 0.1–0.25 mm, 4: 0.5–1.0 mm.

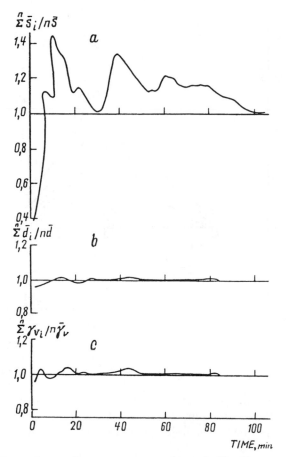

FIGURE 96   Dependence of mean parameters shown in Fig. 64 on averaging time. a) concentration; b) mean diameter; c) coefficient of composition variation.

zone (Kos'yan, 1985a,b). Observations were carried out when a relatively constant swell was acting on the shore. The only plunging area was situated about 32 m from the shoreline, with the breaking zone approximately 12 m long, after which the waves transformed into a bore and migrated shoreward. Using poles equipped with suspended sediment traps and installed at 10 positions in the studied area, a series of observations were carried out, including four exposures for 1 to 1.5 hours each. The location of the measuring posts along the underwater profile is shown in Fig. 97. The posts 1 to 5 were located in the bore zone, posts 6 to 9 were in the plunging zone, and post 10 was seaward of the breaking zone. Stability of the geometric borders of the various zones was monitored visually, while invariability of sediment shift conditions was

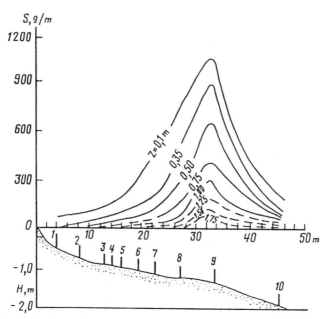

FIGURE 97 Variation of suspended sediment concentration at various levels in breaking and in bore zones.

checked by comparing the results of repeated measurements. During observations, some suspended sediment traps became exposed in the wave troughs (the results of their measurements are shown with the dashed line). In the course of the experiment a sample of surface bed sediment was taken at each point with a special dredge.

The most interesting results of this experiment are noted below. In the plunging zone the mass of suspension sharply increased (in our case 5 or 6 times) as compared with the adjacent area where the sea bottom was influenced by unbroken waves. Concentration values reached their maximum 2 to 3 m shoreward of the point where the plunging began. Over this short distance, the mass of suspension increased from 2 to 3 times. Then there was a rapid, though less sharp, decrease in the concentration up to the zone boundary.

In the bore zone, approaching the shoreline, the suspension content in the water monotonically decreased. The upper layers were clarified more intensively than the lower ones; the suspended sediment concentration decreased by 8 times at 10 cm from the bottom and by 20 times at 50 cm as the shore was approached.

The vertical distribution of the suspended sediment mass in all areas of the breaking zone and bore zone (points 1 to 9) can be approximated by exponential dependence (Fig. 98). This indicates an absence of significant vertical

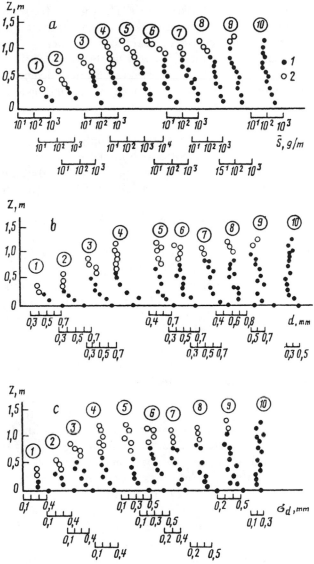

FIGURE 98   Vertical distribution of a) concentration of suspended sediment; b) root-mean-square deviation from average diameter of particles over various points on the underwater slope. Figures in circles show the numbers of measuring posts. 1: traps submerged into water during the whole period; 2: traps exposed in the wave trough.

variations in sediment diffusivity that is possible only during very intensive water exchange throughout the water column. This mixing is caused primarily by the vortices formed in broken waves. The special observations showed that bed microforms wear away and the grain roughness itself is low.

Analysis of distribution curves of grain size parameters shows that the mean size of suspended sediment, as well as the value of the root-mean-square deviation, s_d varies with concentration. Namely, a significant peak is observed in the plunging zone (with the mode corresponding to the maximal concentration point) and a slow decrease in the surf zone as the shoreline is approached.

A decrease of $\bar{d}(z)$ and $s_d(z)$ values occurs mainly in the 20 cm thick near-bottom layer upwards from the bottom throughout the underwater profile. Above the near-bottom layer, in the plunging zone and in the zone of unbroken waves, their distribution is rather uniform, while in the bore zone it slightly decreases.

The mean diameter of bed sediment ($\bar{d}_B$) increases greatly in the plunging zone due to fine fraction washout, and thus it becomes considerably larger than that in neighboring areas of the coastal slope. At point 9, $\bar{d}_B$ reaches the value 0.85 mm while at point 10 (zone of unbroken wave action), $\bar{d}_B$ is 0.45 mm. The mean bed sediment diameter in the bore zone is fairly constant throughout the zone and has a value of 0.60 to 0.65 mm. At the same time, changes in $\bar{d}_B$ values along the profile during calm weather vary insignificantly from 0.65 to 0.66 mm.

Calculated coefficients of asymmetry and excess of suspended and bottom sediment appeared to be close to zero at all levels in the plunging zone and up to the bore zone. In the bore zone and in the zone of unbroken waves, however, they grew significantly from a zero at bottom to a few units on the surface. This means that in the zone of intensive mixing, the distribution of suspended and bottom sediment is log-normal. At the sites with slight mixing, such distributions are observed only at the bottom; the farther from the bottom, the sharper the distribution curve becomes.

It should be added that at an hour's exposure, the "smearing" effect of the plunging line is evident. The degree of its manifestation depends on the wave structure. As the waves in our experiment were close to monochromatic, all the above features were evident. If observations during a single wave were possible, they would be more prominent, while under wind waves these features would be somewhat smoothed.

Water mass mixing results from water orbital motion and turbulence caused by flow interaction with a rough bottom and by surface wave breaking. In the studied zone the orbital motion is transformed and becomes insignificant at its shoreward border. As the bedforms are wiped off here, near-bottom turbidity is low and involves only the thin near-bottom layer. Thus, wave breaking becomes the main factor controlling the suspension distribution throughout the water column. Kana (1978) showed (Figs. 99 and 100) that when the waves are breaking by plunging, flow saturation by sediment is several times greater than when they are breaking by spilling.

6.5 Calculation of suspended sediment concentration and mean diameter

6.5.1 The shoaling zone

It has been shown in Chapter 2 that the model suggested by Kos'yan and Pakhomov (1981) (equation (2.67)) gives a good agreement with the concentrations measured in laboratory conditions. When tested by the results obtained in the underwater laboratory "Chernomor," calculated and measured concentration values were in good agreement (Kos'yan, 1985) when the bottom sediment was suspended by swell waves approximating the conditions of monochromatic waves.

In contrast to swell, wind waves are always irregular and suspension concentration and composition can vary with each wave. To estimate them quantitatively, effective values of wave parameters should be known. Observations of this type were carried out on the North African Mediterranean coast. The comparison of the measured concentration profiles with those computed from (2.67) showed good agreement when the model operated with the wave parameters of recurrence within 13 to 18% (15% in average). This value closely corresponds to the recurrence of the "significant" waves (Sverdrup and Munk, 1947).

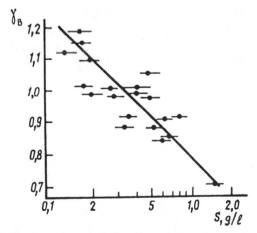

FIGURE 99   Distribution of suspended sediment concentration in the breaking zone (Kana, 1978).

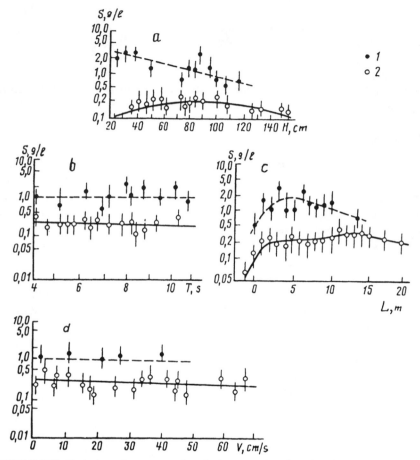

FIGURE 100 Variation of concentration during wave breaking by plunging (1) and by spilling (2) as a function of wave height (a), period (b), distance to the wave breaking point (c), longshore current velocity (d) (Kana, 1978).

6.5.2 The zone of breaking waves

It has been shown above that wave breaking has a major share in flow turbidity and the mixing of water masses and solid particles.

Measurements obtained in field conditions served as a basis for an empirical equation for computation of the relative vertical distribution of suspended sediment concentration in the breaking zone. From the entire data array on the suspended sediment concentration profiles during storms, only those experiments were chosen in which the plunging wave breaking measurements were accompanied by observations on sediment composition, breaking wave height (H_B), water depth at the breaking point (h_B), and visual estimates of the number of waves broken at the observation point (in percent). In this manner, 40 profiles were selected. The measurements were made at depths from 67 to 330 cm; 6 to 14 samples of suspended sediment were collected at each vertical, depending on water depth. The samples were dried and subjected to 16-fraction sieving.

The grain-size analysis showed that in all the samples, seven particle fractions dominated: $d_i = 0.143$; 0.175; 0.225; 0.283; 0.408; 0.65; 0.9; where d_i is the mean diameter of the i-fraction in mm. The mean density of sediment particles in all the experiments was $\rho_s = 2.65$ g/cm^3, and the water density was $\rho = 1.03$ g/cm^3.

In each experiment a separate relative distribution of suspended sediment was plotted for all the fractions. Settling velocities ω_{si} were calculated for all d_i values by the Swart (1976) formula:

$$\log(1/\omega_{si}) = 0.447(\log d_i)^2 + 1.96 \log d_i + 2.736 \qquad (6.1)$$

where d_i is in m, and ω_{si} in m/s.

The value of γ_B can be found with the use of the Rayleigh distribution from the relative number of waves breaking at the observation point:

$$\gamma_B = \frac{H_B}{h_B} \qquad (6.2)$$

The value of γ_B varied from 0.33 to 0.80, i.e., more than two times in accordance with the relative number of breaking waves. Reliable instrumental measurements of γ_B values were not possible in our experiments, and it was very difficult to evaluate them visually at various points of the underwater slope. Thus, the concentration profiles selected for the analysis were all divided into three groups. The first one incorporated the data obtained in the breaking zone of particularly large waves of approximately 1% recurrence during the given storm. The mean value of the γ_B coefficient for this group was about 0.33. The second group corresponded to the visually prominent zone of "significant" wave breaking with approximately 15 to 17% recurrence and the mean value of $\gamma_B = 0.48$. Data obtained on the underwater slope section where 90% of the waves were breaking composed the third group with $\gamma_B = 0.78$.

It has been experimentally proven more than once that the vertical distribution of the suspended sediment concentration in the breaking zone can be approximated by exponential dependence which indicates the constancy of diffusion coefficient values with flow depth (Kos'yan, Pykhov, and Phylippov, 1978; Kos'yan, Dachev, and Pykhov, 1980; Kos'yan, 1985). Thus, in accordance with (2.41), the profile of the concentration averaged for a measuring time can be written as

$$\frac{\overline{S}(z)}{\overline{S}_C} = \exp(-K_b z) \tag{6.3}$$

where K_b is the gradient of the logarithm of concentration.

For specific computations of concentration profiles by (6.3), K_b values for the given conditions should be set. Concentration curves for narrow fractions based on observations of the suspended particle distribution in the water column, are shown in Fig. 101. The value of K_b for each of them can be found from (6.3):

$$K_b = -\frac{d(\ln \overline{S}(z))}{dz} \tag{6.4}$$

The results of the previous experimental research suggest that the gradient of suspended sediment concentration in the breaking zone depends on the particle settling velocity (Kos'yan, Dachev, and Pykhov, 1980), wave height, water depth (Antsyferov and Kos'yan, 1986), and the relative number of the waves breaking at this point (Kos'yan and Kochergin, 1987). Thus, K_b can be presented as a functional dependence:

$$K_b h_B = f(\gamma_B, h_B \omega_s / v) \tag{6.5}$$

Concentration profiles obtained in the wave breaking zone of 1% probability (Fig. 101) are closer to those of the zone of unbroken wave action (Kos'yan, Dachev, and Pykhov, 1980) than to the profiles in the breaking zone. The sediment suspension outside the breaking zone and in the zone where the bottom is affected only by a few breaking waves seems to be controlled mainly by the turbulence of the near-bottom boundary layer and orbital water motion. For this reason the group of K_b values corresponding to $\gamma_B = 0.33$ is excluded from further analysis.

The remaining data on concentration gradient measurements, presented in Table 12 have been used to find the empirical relationship of K_b in the form:

$$K_b h_B = 0.19(1 - 0.33/\gamma_B)^{1.7}(h_B \omega_s / v)^{0.58(0.33/\gamma_B)^{0.56}} \tag{6.6}$$

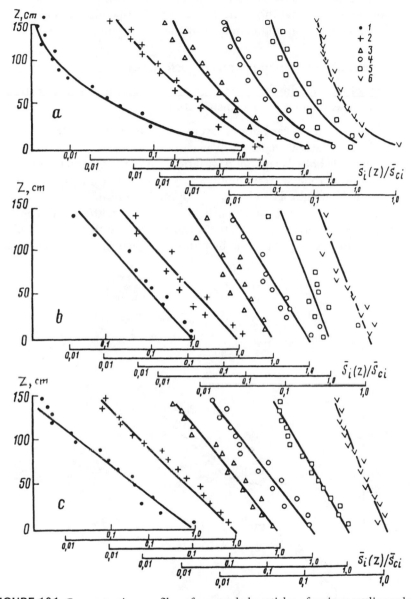

FIGURE 101 Concentration profiles of suspended particles of various settling velocities at: a) $\gamma_B = 0.33$; b) $\gamma_B = 0.48$; c) $\gamma_B = 0.78$. S_{ci} is the extrapolated value of i-fraction concentration at bottom. 1: $\omega_s = 9.5$ cm/s; 2: $= 6.0$ cm/s; 3: $= 3.9$ cm/s; 4: $= 2.7$ cm/s; 5: $= 2.2$ cm/s; 6: $= 1.4$ cm/s.

The measured values of K_b compared with those computed by equation (6.6) show (Fig. 102) that the vertical gradient of suspended sediment concentration is satisfactorily approximated by this expression.

6.5.3 Vertical distribution of the mean particle size

Analysis of the experimental data suggests the following characteristics of the suspended sediment mean size distribution:

1. In the 20-cm thick near-bottom layer, $\overline{d}(z)$ values decrease significantly upwards from the bottom nearly in all cases.
2. Above the near-bottom layer, $\overline{d}(z)$ variations do not exceed 10 to 20%.
3. In the breaking zone distribution of $\overline{d}(z)$ is more uniform than in the zone of deformed but unbroken wave action upon the bottom.
4. The mean size of suspended particles $\overline{d}(z)$ and bottom sediment $\overline{d}_B$ increase with the deformation of the surface waves and the maximal values coincide with the wave breaking by plunging, when the $\overline{d}(z)$ profiles vary insignificantly.

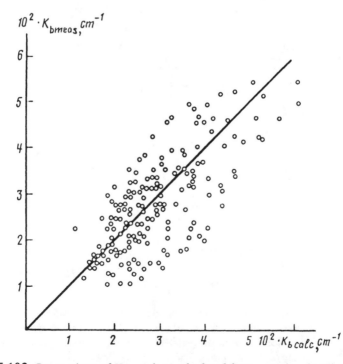

FIGURE 102 Comparison of K_{bcalc} values calculated from equation (6.13) with K_{bmeas} values measured during field observations.

Accounting for the characteristics of the vertical distribution of $\overline{d}(z)$, and on the basis of natural observation data, an empirical relationship for $d_*(z) = \overline{d}(z)/\overline{d}{>}_B$ may be found. As yet, this relationship cannot be referred to all the appropriate values (wave parameters, bottom sediment composition, the degree of wave deformation), so the vertical profiles of the suspended sediment mean diameter beneath the unbroken and breaking waves, averaged by all the measurements, are considered. 155 profiles of $\overline{d}(z)$ from the unbroken wave zone and 30 profiles from the breaking zone were selected from the Kamchia testing site measurements.

From the $\overline{d}_*(z)$ variations averaged by Δz intervals in all the profiles, the curves of the most probable $\overline{d}_*(z)$ values and ranges (Fig. 103) are obtained. Averaged vertical profiles and ranges of sediment composition variability coefficients γ_v/γ_{vB} (where $\gamma_v = \sigma_d/\overline{d}$, $\gamma_{vB} = \sigma_d/d_B$. This figure shows that with distance from the bottom suspended sediment grading varies less than the mean particle diameter.

The averaged curves of $\overline{d}_*(z)$, shown in Fig. 103, can be approximated by the following empirical formulas:

For the unbroken wave zone $\qquad \overline{d}_*(z) = \exp\left[-0.054(z)^{0.5}\right]$ $\qquad$ (6.7)

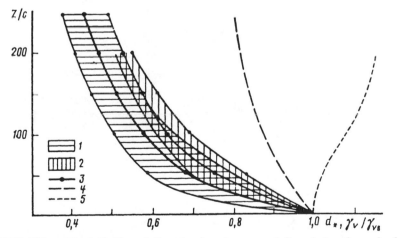

FIGURE 103 Vertical profiles of d^* and γ_v/γ_{vB}. 1: range of d^* variation for unbroken waves; 2: the same for breaking waves; 3: most probable value of d^* for unbroken waves (equation (6.7)) and for breaking waves (equation (6.8)); 4: averaged profiles of γ_v/γ_{vB} for unbroken waves; 5: the same for breaking waves.

For the breaking zone $\qquad \bar{d}_* (z) = \exp\left[-0.032(z/c)^{4/7} \right] \qquad$ (6.8)

where $c = 1$ cm.

These equations are based on the measurement results on a single section of the Black Sea coast, but the comparison of the computations and the measured values obtained on the Baltic Sea and Mediterranean coasts and other sections of the Black Sea coast [Antsyferov and Kos'yan (1986); Onischenko and Kos'yan (1989)] suggests their applicability for practical estimates in other regions (Fig. 104).

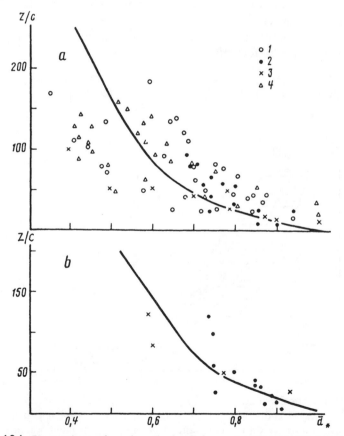

FIGURE 104 Comparison of results calculated from equations (6.7)(a) and (6.8)(b) with the measured values from the Mediterranean (1), Baltic (2), and Black (3,4) Seas.

6.6 Prediction of suspended sediment distribution over the entire nearshore zone

Prediction of the suspended sediment concentration distribution in a region with varying bottom sediment conditions presents the greatest interest. Under certain conditions the problem of prediction reduces to the calculation of concentration above the areas typical in relief, sediment composition, and dynamic environment under given parameters of surface waves with the subsequent interpolation throughout the water area. The models for calculating the mean concentration vertical profiles $\overline{S}(z)/\overline{S}_c$ discussed here and in Chapter 2, together with the recommendations on $\overline{S}_c$ calculation, enable one to determine the total amount of the suspended sediment in a water column.

Taking into account the hypothetical nature of $\overline{S}(z)/\overline{S}_c$ extrapolation, concentration regime functions based on wave regime functions can be constructed for several levels and for depth-averaged values.

Having constructed concentration regime functions of suspended particles for all areas above the sand floor, and taking into account the clarification of sediment flow above the rock floor, the distribution of suspended particles can be predicted in the whole water area under study for any direction and return period. The calculation results for depth-averaged concentrations of suspended sediment (the North African shelf section) are presented in Kos'yan (1984c), Antsyferov and Kos'yan (1986), and in Chapter 7 of this volume (for the Bulgarian Black Sea coast).

PART

III

Sediment Transport

CHAPTER

7

Longshore Sediment Transport

The ultimate purpose of sediment transport studies is the prediction of bottom relief in the zone of active wave effect accompanied by the transport of significant sand volumes. The modeling of bottom relief variations is based on the equation obtained from bottom to surface integration of the conservation of mass:

$$(1-n)\frac{\partial h}{\partial t} = \frac{\partial Q_x}{\partial x} + \frac{\partial Q_y}{\partial y} \tag{7.1}$$

where $Q_x(x,y)$ and $Q_y(x,y)$ are the depth-averaged cross-shore and longshore components of sediment discharge per bottom unit area; n is a coefficient of the sediment porosity; the axis x is directed seaward, and the axis y is directed along the shore. It follows from (7.1) that to predict variations in morphology, sediment discharge at each point (x,y) should be known in the chosen time scale. This scale can be of hours, if the detailed relief variations during a single storm are to be registered, or of a year, if the multi-year variations on the selected coastal area are to be studied.

Sediment discharge components Q_x and Q_y can be computed from the known concentration and velocity field distributions:

$$Q_x(x,y) = \int_{-h}^{H} (1/T_m)\int_0^{T_m} S(x,y,z,t)V_x(x,y,z,t)dtdz \tag{7.2}$$

$$Q_y(x,y) = \int_{-h}^{H} (1/T_m)\int_0^{T_m} S(x,y,z,t)V_y(x,y,z,t)dtdz \tag{7.3}$$

209

where $S(x,y,z,t)$ is the sediment volume concentration, $V_x(x,y,z,t)$ and $V_y(x,y,z,t)$ are the cross-shore and longshore components of the sediment transport velocity, and T_m is the selected averaging time.

The current difficulties in determining the concentration and velocities have been mentioned previously in the analysis of elementary processes and measuring data. Thus, the rigorous prediction of sediment discharge under the actual conditions by (7.2) and (7.3) seems impossible, though this approach can be used for rough evaluations in the simple cases.

The energetic longshore currents induced by the approach of oblique waves, transport large amounts of sand lifted by waves from the sea bottom. This mass sediment transport and its longshore variations finally determine the shoreline configuration and the location of accretion and erosive areas on the underwater slope. Because of the practical significance, the problem of longshore sediment transport has attracted much attention.

7.1 In situ measurements

In situ measurements of sediment discharge are currently rather limited due to significant technical and methodical difficulties. Sediment discharge in field conditions has been measured by the seasonal variations of the bottom relief, or by the results of tracer experiments during certain storms, until Greer and Madsen (1978) convincingly proved them to be worthless for the correct evaluation of longshore sediment transport. Attempts of direct measurement of sediment discharge are few (Aibulatov, 1966; Allen, 1985; Voitsekhovich, 1986a,b); their advantages and disadvantages being discussed by Leont'ev (1988a). The results of the measurements at the Kamchia testing site (Leont'ev and Pykhov, 1988), visually representing the structure of the suspended sediment longshore discharge, will be discussed.

The longshore sediment transport was studied in 1983 and 1985 at the "Shkorpilovtzy" research site of the Bulgarian Academy of Sciences. During these experiments, wave parameters, flow velocities, and suspended sediment concentrations in the water column were recorded synchronously at several points located along a range line of the research pier. In accordance with the wave strength, observations were made either along the entire length of the pier or on its shoreward section. The typical bottom profile and measuring point locations are shown in Fig. 105. Water flow velocities were measured at 3 to 4 levels from the bottom to the surface, the lowest level being 0.1 m above the floor. The duration of the recording was 10 min.

Sediment concentrations were measured with the suspended sediment traps distributed in the lower 1.5 m water layer at an interval of 0.1 m (the lowest trap being 0.1 m above the bottom). The usual exposure time was 1 hour.

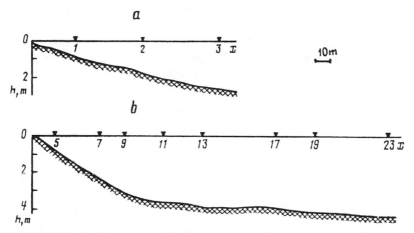

FIGURE 105 Typical bottom profiles and location of measuring posts for run A.-2 and A.-3(a); and A-4–A-8(b).

The elementary longshore suspension discharge per unit flow section area can be computed from the measured concentration and velocity values:

$$q'_s = \bar{S}(z)\overline{V}(z) \tag{7.4}$$

To obtain the necessary velocity values between the measuring levels, linear interpolation was adopted. The subsequent integration of q'_s values with respect to layer height at the given vertical gave the value of transport velocity or specific discharge of suspension (per unit of longshore current width):

$$q_s = \int_{z_{min}}^{z_{max}} \bar{S}(z)\overline{V}(z)dz \tag{7.5}$$

where $z_{min} = 0.1$ m; and $z_{max} = 1.5$ m or was equal to depth if the latter exceeded the 1.5 m. The measurements showed that the sediment in the layer above 1.5 m was practically negligible.

Finally, integration of q_s with respect to the profile length (i.e., longshore current width) gave the total longshore discharge of suspension:

$$Q'_s = \int_0^{x_B} q_s dx \tag{7.6}$$

where x_B is the longshore current width. The value of q_s at the water edge was assumed to be equal to zero. It should be remembered that q'_s, q_s, and Q'_s values correspond to the dry weight of suspended sand transported in a unit of time.

Mean wave parameters and their frequency energy spectrum were estimated through observations. The depth of wave breaking initiation was found from the root-mean-square wave height H_{Brms} under the condition:

$$\frac{H_{Brms}}{h_B} = 0.37 \tag{7.7}$$

The angle of the wave approach was evaluated visually.

A total of 5 runs of longshore discharge measurements (designated A-4 to A-8) were obtained during the experimental studies in 1985. In runs A-6 through A-8, observations were made at points 5, 9, 13, 17 (or 19), and 23 along the entire range line of the pier, while in runs A-4 and A-5 at points 5, 7, 9, 11, and 13 (Fig. 105b). During the experiments of 1983, two observation series were made along a rather short range line (A$_*$-2 and A$_*$-3, Fig. 105a).

The general wave characteristics are given in Table 13, where T is the period corresponding to the main peak of the wave energy spectrum; H_{Brms} and T coincide with the beginning of the range line.

Nearly all the measurements were made under swell, with the exception of run A$_*$-3 which was performed during wind waves. The range lines were normal to shore being directed from east to west. In all the series, waves approached the range lines from the north and induced southern longshore transport which will hereafter be considered as positive.

TABLE 13   Characteristics of the suspended sediment longshore transport

Parameters	Run						
	A*–2	A*–3	A–4	A–5	A–6	A–7	A–8
H_{brms}, m	0.61	0.74	0.68	1.23	1.55	1.41	1.58
T, s	5.0	5.8	5.1	5.7	7.3	8.3	7.2
h_B, m	2.0	2.4	1.9	3.4	4.2	3.9	4.2
α_B	30	30	30	30	10	10	10
Q_s', kg/s	3	22	6	30	20	21	57
$\bar{d}$, mm	0.4	0.3	0.4	0.4	0.4	0.4	0.3
$_s\omega_s$, cm/s	6.0	4.0	6.0	6.0	6.0	6.0	4.0
$(gh)_B^{1/2}/\omega_s$	74	121	72	96	107	103	160
$F_{yB} \cdot 102$, N/s	15	23	13	56	33	27	33
$c_s \cdot 10^{-2}$	0.5	1.2	1.0	0.9	0.9	1.2	1.7
Q_s^T, N/s	42	108	36	204	134	108	200

The important experimental results are shown graphically in the accompanying figures. The distributions along the profile of longshore velocities and suspension concentrations averaged by depth and designated as $\overline{V}$ and $\overline{S}$, respectively, are shown in Fig. 106 together with the specific discharges of the suspended sediment. The typical curves of longshore velocities are shown in Fig. 107 for several observation points. More detailed data on the vertical structure of the suspension flow can be seen in Figs. 108, 109, 110, and 111.

Fig. 106 shows that the flow velocities were on the order of tens of centimeters per second. The maximum values in most of the series were approximately 0.5 m/s, while in run A-8 they exceeded 1 m/s.

The velocity distribution along the profile has one or two peaks. The unimodal distribution is observed in the cases of slight agitation (runs A.-2 through A-5), the velocity maximum being registered at the depth of approximately 0.5 m in the point nearest to shore. The actual position of the maxi-

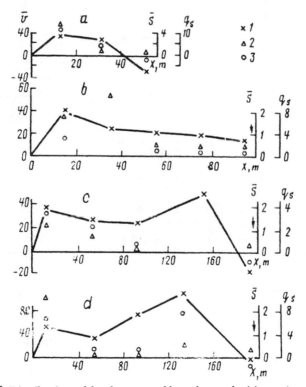

FIGURE 106 Distribution of depth-averaged longshore velocities, m/s; concentration of suspended sediment, 10-3g/cm³ (2) and suspended sediment rate, g/cm s (3) in runs A.-3(a), A-5(b), A-6(c), A-8(d). (The arrows show the breaking depth of the waves of 1% recurrence.)

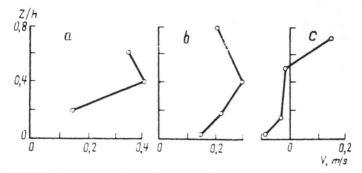

FIGURE 107 Vertical curves of longshore velocities at points 5(a), 13(b), and 23(c).

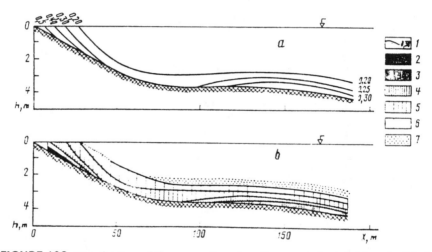

FIGURE 108 Distribution of the mean diameter of suspended sand (a) and of their concentration (b) along the underwater profile in run A-8. 1: isolines of equal mean diameter of suspended sediment; 2–7 concentration values: $>5.10^{-3}$; 1.0–5.0; 0.5–1.0; 0.1–0.5; 0.05–0.1; $0.05 \cdot 10^{-3}$ g/cm³, respectively.

mum can evidently be even closer to the shore as the intensive longshore transport was observed in the entire area adjacent to the shore. Under higher wave activity this velocity peak remained in the same zone adjacent to the shoreline, while the second, more prominent peak appeared in the zone where large waves were breaking. Here the maximum velocities for the entire range line were observed.

A noteworthy feature of the velocity distribution is the reversal of the flow direction near the seaward limit of the surf zone (at the depth of h_B). This phenomenon can be traced both along the short range lines of 1983, and along those of runs A-6 through A-8. The longshore velocity reversal seems to

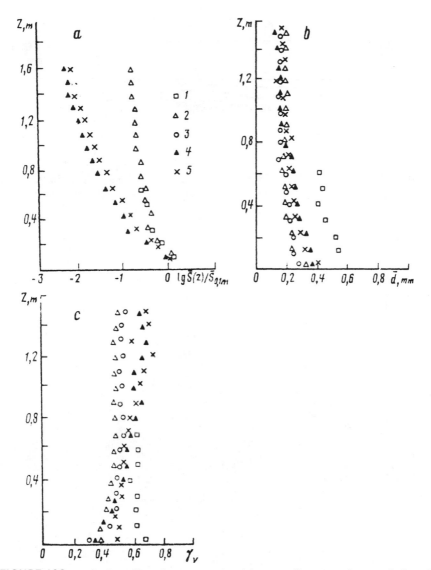

FIGURE 109 Vertical profiles of concentration (a), mean diameter of suspended sand (b), and coefficient of composition variation (c). 1–5: points 5, 9, 13, 17, 23 respectively.

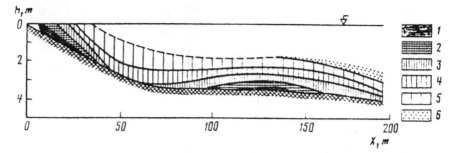

FIGURE 110 Distribution of elementary discharges of suspended sand in the lower 1.5 mm layer of the water column in run A-8. 1-6: values of $q_s > 0.1$; 0.05–0.1; 0.01–0.05; 0.005–0.01; 0.001–0.005; <0.001 g/cm²/s.

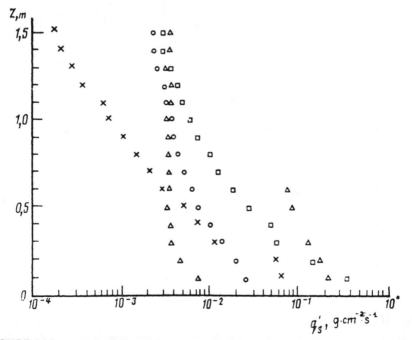

FIGURE 111 Vertical profiles of elementary discharges. (For specifications see Fig. 109.)

be caused by nearshore circulation (of the circulation cell type) induced by the non-uniform bathymetry and shoreline. Underwater relief variations are known to result in non-uniform distribution of the mean longshore water level. This level becomes higher in shallower areas, and gradient currents directed toward deeper areas are generated (Leont'ev, 1987). Thus, if the velocity reversal is related to the gradient current, general subsidence of the sea floor to the north of the measuring range lines can be expected. This tendency was actually observed along the range line of the 1983 experiments, which bordered a vast shoal on the south. In 1985 similar observations along the range line were not made.

In general, gradient currents are weaker than the energy currents induced by the oblique wave approach, and primarily manifest themselves mainly beyond the surf zone (Fig. 106). Only in run A.-2 are both current types comparable in intensity.

The velocity profiles in the zone of longshore current reversal are of especially particular form (Fig. 107). The two layers are clearly seen; in the upper one the transport is controlled by the energy current, while in the lower one, the reverse gradient current takes precedence. Sometimes the gradient current strengthens toward the bottom, while in other cases its maximum coincides with the higher levels.

The energy current is also non-uniform in the vertical section (Fig. 107a, b) and velocity profiles vary in form. The velocities seem to decrease towards the bottom.

With suspended sediment, the $\bar{S}$ distributions tend to contrast. Along the major part of the profile, $\bar{S}$ values are low (about 10^{-4} g/cm^3), while close to shore they increase greatly with the maximum values sometimes amounting to 10^{-2} g/cm^3. The appearance of the second $\bar{S}$ and $\bar{V}$ peaks does not always correlate. Thus, the second $\bar{S}$ peak is prominent in run A-8, but is absent in run A-7. This possibly results from the different conditions of wave breaking, implying their nonuniform action upon the bottom.

The concentration distributions and the mean size $\bar{d}$ of the suspended sediment in the water column and along the profile (series A-8, moderate agitation) are shown in Fig. 108. The similarity of the $\bar{S}$ and $\bar{d}$ distributions is clearly seen. The increased values of both parameters correspond to areas of wave breaking. The thickness of the suspension layer is also maximum here.

The vertical profiles of $\bar{S}$ and $\bar{d}$ at various points of the underwater slope are shown in Fig. 109 for the run A-8. For convenience of comparison, the concentration values are normalized to the $\bar{S}$ value at the 0.1 m level. Minimum vertical gradients of the concentration are observed in the wave breaking zone (points 5 and 17), where its value varied by an order of magnitude, while at other points it varies by two orders of magnitude.

In contrast to concentration, the value of $\bar{d}$ varies insignificantly with distance from the sea floor (Fig. 109b), point 5 being the only exception where $\bar{d}$

decreases rapidly from the bottom to the approximately 0.3 m. This is explained by the increased amount of coarse-grained particles which cannot penetrate far into the water column when suspended. At other points, vertical variations do not exceed 100%.

Cross-shore distributions of the specific sediment discharges are shown in Fig. 106. The variations of q_s generally follow the concentration variations. The highest values of q_s (up to 10 g/cm s) are generally observed close to the shoreline where the concentration peaks. Bimodal distributions are observed only in run A-8. The share of the longshore current reversal in the total suspension transport is insignificant due to low values of $\overline{S}$.

The distributions of elementary discharges q_s' in the water column for run A-8 are shown in Fig. 110. Similarity to the concentration distribution is evident; the wave breaking areas play the major role in sediment transport.

The vertical structure of the suspension discharge is illustrated by Fig. 111. It can be seen that the more intensive transport is also more uniform in the vertical direction (it varies within an order of magnitude). Outside the wave breaking zone, the leading role in sediment discharge belongs to the near-bottom layer up to 0.5 to 0.7 m thick.

The total sediment discharges Q_s' are presented in Table 14. The figures visually demonstrate the immense masses of the suspended sediment transported alongshore during storms. In the A-8 series, for example, more than 200 tons of suspended sediment apparently passed through the measuring range line during one hour.

These data characterize the suspension discharge in the main flow column, while the lowermost 10-cm thick layer is ignored due to the inability to sufficiently measure water velocity and sediment concentration in this region. The latter evidently has some share in the suspension discharge, though its exact role can only be hypothesized.

It can be assumed that the suspension transport velocity within the named layer is of the same order as that at the 0.1 m level. This is quite possible if the particle concentration is roughly uniform here due to intensive turbulent exchange. In this case, the additional suspension discharge produced by the near-bottom layer does not exceed 10% which lies within the measurement error. As the data obtained in this manner actually characterize the total suspension discharge, the bedload discharge can be found. The estimates show that for the studied conditions, the bedload discharge Q_B should be 0.6 to 1.4, (i.e., the discharges of bedload and suspended sediment are of the same order). Therefore, the total discharge should exceed the measured suspension discharge by 1.6 to 2.4 times.

The share of the near-bottom layer in the sediment transport can be estimated from another point of view. The particle concentration is assumed to increase continuously in this layer to some threshold value at the bottom where the transport velocity tends to zero. Thus, if $\overline{S}$ grows exponentially up to the threshold value of 0.25 g/cm^3 and $\overline{V}$ decreases linearly to zero, the near-bottom sediment discharge is of the same order as in the above-water column. Under this approach, bedload and suspended sediments are not differentiated. Nevertheless, the resulting summary discharge is roughly similar to that obtained in the first case.

7.2 Semi-empirical models

The results of the field observations described above reveal the complexity of the spatial structure of suspended sediment discharge along the shore. Due to the almost insurmountable difficulty in directly measuring the bedload sediment transport, it is impossible to make accurate predictions. On the other hand, the immense practical significance of this problem stimulated the development of some tens of equations for longshore sediment transport evaluation during the last 20 years. Various aspects of this problem have been reviewed in Longinov (1966); Voitsekhovich (1986b); Beloshapkov (1988a,b); Phylippov (1988); Hallermeier (1982); and the joint monograph *Nearshore Dynamics and Coastal Processes* (Horikawa, 1988). Omitting all the concepts on which the suggested equations are based, let us turn to the most promising among them. In the most numerous group of studies, the discharge value averaged over the surf zone cross section is assumed to be proportional either to dimensionless combinations of wave parameters, bottom slope, bed sediment composition, or to the longshore flow of the wave energy calculated from the wave parameters along the breaking line (Rybak and Suprunov, 1982; Voitsckhovich, 1986b; Beloshapkov, 1988; *Shore Protection Manual*, 1984; Komar, 1977; Kamphuis and Rcadshaw, 1978; Ozasa and Brampton, 1980).

The suggested equations can all be reduced to the form:

$$Q_y = A_y F_{yB} \tag{7.8}$$

$$F_{yB} = (Ec_g)_B \cos\alpha_B \sin\alpha_B \tag{7.9}$$

where E is energy; c_g is the group velocity; α_B is the angle of wave approach relative to the normal at the wave breaking line. The integral discharge Q_y is expressed in the units of submerged sediment weight. Among the equations of this group, the formula suggested by the U.S. Army Corps of Engineers, Waterways Experiment Station (*Shore Protection Manual*, 1984) was most frequently used :

$$Q_y = A_y (H_{rms}^2 c_g)_B \sin 2\alpha_B \tag{7.10}$$

$$A_y = \frac{k_1}{16(\rho_s/\rho-1)(1-n)(1.416)^{5/2}}$$

The value of $k_1 = 0.77$ was originally determined by Komar and Inman (1970) from tracer experiments. The replicate analysis of the same data (Greer and Madsen, 1978) later showed that they cannot be used for correct evaluation of k_1. Kraus et al. (1982) recommended $k_1 = 0.58$ on the basis of their tracer experiments.

Evaluation of A_y using laboratory measurements (Kamphuis and Redshaw, 1978; Hallermeier, 1982; Ozhan, 1983) showed that this coefficient depends on the type of wave breaking defined by the Battjes parameter:

$$\xi = \frac{\tan\beta}{(H_o/\lambda_o)^{\frac{1}{2}}} \tag{7.11}$$

where H_0, λ_0 are the height and length of monochromatic waves in deep water. Analyzing the laboratory measurement data published by 1981, Haller-meier (1982) found $A_y = 0.11$ for $0.17 < \xi < 0.41$, $A_y = 0.17$ for $0.42 < \xi < 0.52$, respectively. These are the typical values for waves breaking by spilling. For $0.52 < \xi < 1.01$, typical of plunging waves, $A_y = 0.32$ was obtained. These data indicate greater longshore sediment transport due to wave plunging, which is in agreement with the growth of the suspension concentration for this type of wave breaking (see Chapter 6).

Among the Russian studies of this group, the equation derived by Beloshapkov (1988) from (7.8) and (7.9) using solitary wave theory for calculating the longshore flow of wave energy and based on all the laboratory and experimental data for evaluation of the empirical coefficients, should be noted:

$$Q_y = 0.075\rho g \left(\frac{H_{sB}^2}{T_B}\right)\left(\frac{H_{sB}}{h_B}\right)^{-3/2}\left(\frac{\bar{d}}{d_0}\right)^{-1/2}\sin\alpha_B \tag{7.12}$$

where T_B is the wave mean period, $d_0 = 1$ mm, and H_{sB} is the significant wave height. Comparison of Q_y values calculated by (7.12) with those measured in all the laboratory and in situ studies showed that the possible calculation errors can amount to 50%.

Japanese scientists widely use the model suggested by Ozasa and Brampton (1980), allowing both for the longshore flow of wave energy and for a long-shore gradient of wave heights. This factor should be accounted for when calculations concern breakwaters and other hydraulic works, as the currents originating on their shadow side are directed toward them because of lower wave heights due to diffraction (Horikawa, 1988).

More recently the energy concept of Bagnold has been widely used for evaluation of longshore sediment discharge (Bagnold, 1963, 1966, 1973, 1979). According to this concept, the flow power spent on particle displacement in a unit liquid column (expressed through particle concentration and the rate of their transport) is proportional to the power lost on interaction with the bottom. In a number of studies (Bailard, 1981; Leont'ev, 1988) this concept has been adopted to longshore currents with superimposed wave oscillations. It is assumed in these models that at greater near-bottom oscillatory velocities, the sediments are in constant motion, so that Bagnold's relationships can be used in which depth-averaged velocity is replaced by the longshore current and orbital velocities. In Bailard's model (Bailard, 1981), the orbital velocity field is given by the Stokes' second order approximation, while the longshore current is described by the model suggested by Ostendorf and Madsen (1979). According to Bailard's model, defining both local discharge versus the surf zone width and the integral discharge, the majority of the sediment is transported in a band twice as wide as the surf zone, the maximum being located close to the wave breaking line.

In Leont'ev's model (Leont'ev, 1988), the orbital and longshore velocities are described with the help of his model for irregular waves. The resulting equation of the integral discharge versus the surf zone width has the form:

$$Q_y = \left[0.186 \frac{c_B}{\tan\phi} + 0.038 c_s \frac{(g^{1/2}h_B)}{\overline{\omega}_s} \right] F_{yB} \tag{7.13}$$

where c_B and c_s are the efficiency coefficients for bottom and suspended sediment; h_B is the breaking depth of the waves of 1% recurrent; and F_{yB} is also calculated from the 1% recurrence wave parameters. Calibration of this model with field measurement data, discussed in the previous paragraph, gave the range of c_s values from 0.005 to 0.017 (see Table 14), 0.01 on the average, which agrees with Bagnold's estimates for the current flows (Bagnold, 1966). The absence of measurement data makes determination of c_B rather difficult. If c_B for the wave flow is assumed to be equal to that for the steady flow $c_B = 0.1$, then in the typical environment of the nearshore zone the value of A_y in (7.8) would be 0.08 to 0.16 (Leont'ev, 1988).

The attempts at directly estimating discharge by equation (7.2) are very few. This approach has been developed intensively by Dutch scientists. In their models, the channel hydraulic relationships are adapted for the joint action of waves and currents. These models are generally very huge with a large number of empirical coefficients, and thus they cannot be placed above

TABLE 14   The values of the concentration logarithm gradient based on the data of field studies ($K_b \times 10^{-2}$)

h_B, cm		$\gamma_B = 0.48$						$\gamma_B = 0.78$							
ω_s, cm/s	67	125	166	252	276	286	67	78	86	95	112	120	275	330
13.0	5.1	2.2	2.7	–	2.4	2.9	4.1	5.2	3.1	4.6	4.8	3.1	–	–
	4.6	2.9	5.4	–	–	–	–	–	5.1	–	3.6	4.0	–	–
	–	–	3.2	–	–	–	–	–	–	–	3.9	–	–	–
9.5	4.2	2.0	1.7	3.1	2.3	2.6	4.0	4.3	1.9	4.8	4.9	3.0	3.4	1.9
	4.6	2.2	3.8	–	–	–	–	–	4.5	–	3.4	–	–	–
			3.1	–	–	–	–	–	–	–	3.8	–	–	–
			3.7	–	–	–	–	–	–	–	1.7	–	–	–
	3.3	1.6	2.0	2.9	2.1	2.3	3.4	3.4	1.7	3.9	4.5	2.7	2.6	1.6
	4.6	2.0	3.6	–	–	–	–	–	3.9	4.6	2.7	2.6	–	–
6.0	5.4	–	3.8	–	–	–	–	–	–	–	3.6	–	–	–
	4.9	–	3.5	–	–	–	–	–	–	–	1.3	–	–	–
	2.6	–	–	–	–	–	–	–	–	–	–	–	–	–
	4.6	1.4	1.8	2.4	1.7	1.7	2.9	2.9	1.4	3.3	3.3	2.1	1.8	1.6
3.9	4.2	1.9	3.3	–0	–	–	–	–	2.6	4.2	2.6	2.3	–	–
	3.8	–	3.0	–	–	–	–	–	–	–	1.1	–	–	–
	2.3	1.5	1.7	2.4	1.6	1.5	2.7	2.2	1.3	3.7	3.0	2.7	1.8	1.4
	3.5	1.9	2.9	–	–	–	–	–	–	3.2	2.3	2.0	–	–
	3.4	–	2.4	–	–	–	–	–	–	–	3.4	–	–	–
2.7	3.3	–	2.6	–	–	–	–	–	–	–	1.1	–	–	–
	2.7	–	–	–	–	–	–	–	–	–	–	–	–	–
	2.5	1.3	1.8	–	1.3	–	2.4	1.8	1.2	2.9	2.3	2.7	1.4	1.6
	3.5	1.4	3.1	–	–	–	–	–	–	3.2	2.2	1.7	–	–
2.2	2.7	–	2.3	–	–	–	–	–	–	–	3.1	–	–	–
	3.0	–	2.4	–	–	–	–	–	–	–	1.1	–	–	–
1.4	1.4	1.2	1.4	1.4	1.4	–	2.5	1.7	1.0	3.1	2.0	2.2	1.1	–
	3.1	1.2	2.0	–	–	–	–	–	–	2.7	2.2	1.5	–	–
	2.3	–	1.4	–	–	–	–	–	–	–2.8	–	–	–	–
	3.1	–	2.9	–	–	–	–	–	–	–	1.0	–	–	–
	1.4	–	–	–	–	–	–	–	–	–	–	–	–	–

those of the two previous groups. These models are described and reviewed in a number of articles (Bijker, 1971; Van De Graaf and Van Overeem, 1979).

In Kos'yan and Phylippov's study (1990), the discharge is estimated using equation (2.67), concentration profiles in the zone of unbroken waves and in the surf zone are calculated from equations (6.3) and (6.6), respectively, and the velocities of longshore currents are found from the Longuett-Higgins model. The calculation scheme and the results of this model are discussed in the next paragraph. It should be noted as a conclusion that the great number of models is a direct indication of how little is known about the physical basis for these phenomena already mentioned in the first two chapters.

Universal models for predicting the longshore sediment discharge are still lacking. Models in the first two groups have simple structures based on the general physical concepts; calculations based on them primarily depend on the correct specification of the constants involved. The models based on known sediment concentration profiles and transport velocities, have similar disadvantages. Still, the latter approach seems more promising as, in contrast to the energy approach, as our knowledge increases, it will become possible to calculate the three-dimensional structure of sediment discharge with varying settling velocities (see 7.1).

7.3 An example of calculation of suspended sediment longshore discharge

7.3.1 The modular calculation scheme

It is convenient to present the calculation sequence as a modular scheme (Fig. 112) consisting of several interrelated algorithmic operations (Kos'yan and Phylippov, 1987, 1988; Kos'yan, 1993). Step-by-step calculations were made from one block to another by the chosen physical models and methods. This scheme provides for operative alterations and additions to separate modules in compliance with the environment of the studied coast. Besides, the content of the calculation modules can be changed without alternating the calculation sequence within the scheme, as soon as better knowledge of the physical processes is attained.

The observation data on wind direction and velocity, bottom relief, and geology in the studied nearshore zone are transmitted from the coastal stations to the calculation scheme input. The nearshore zone is divided into roughly uniform sections, and the deepwater wave parameters, their transformation and refraction in shallow water, the direction and velocity of the longshore current, suspended sediment concentration, the cross-section of the suspended flow, and, finally, the sought-for value of the suspended sediment discharge are calculated sequentially for each of them.

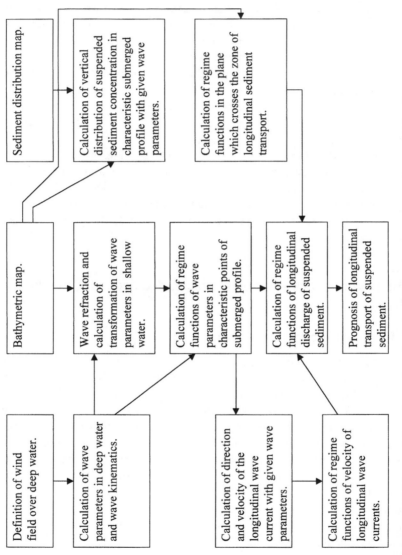

FIGURE 112 The modular scheme for calculation of longshore transport of suspended sediment.

Intermediate and final results are both presented in a probabilistic form as regime functions or recurrence curves. The calculation procedure can be completely automated. The calculations of wind parameters in deepwater areas are incorporated in one of the first blocks. Observations from the nearest coastal hydrometeorological stations, or analysis of synoptic maps over a multi-year period (usually 20 to 25 years) generally serve as the data source. In our calculations, the method suggested in the *Manual of Calculations of the Hydrological Parameters in the Nearshore Zone* (1973) is used, providing for calculating distribution functions of wind strength and direction above deepwater areas.

The "significant" wave parameters in the deepwater areas are calculated in the next block, together with the calculations of wave conditions under the given computed wind velocities and "simple" conditions of wave formation. Functions of wind characteristics are used as input data in this case. Any model satisfactorily describing the process contained in the block can be used for calculations of the suggested scheme. We have chosen the calculation method (*Manual...,*1973) widely used in Russia and recommended for computations of surveying results for hydraulic construction in the coastal zone.

On the basis of the deepsea wave data and a bathymetric map of the chosen coast section, refraction plans and shallow-water wave parameters are calculated next. For this purpose the algorithm developed in the Black Sea Department of CNIIS (Kantarzhi et al., 1983) was used. Regime functions of shallow water wave elements at typical points of the underwater slope are found from the deepsea waves. Then, together with the angles of wave approach and isobaths, they are used in the next block to calculate the longshore current velocities by the Longuett-Higgins model (Longuett-Higgins, 1970), the main elements of which have been described in Chapter 1.

The velocities of the longshore energetic current and the regime functions of shallow water wave parameters serve as source data for the next block, in which the regime functions of longshore current velocities are formed for all wave directions of the studied coast section. The velocity, recurrence, and direction of the resulting longshore current formed by storm-induced currents approaching the shore at different angles relative to the shoreline normal, are needed for further calculations. The recurrence of the resulting current velocities is a sum of velocity recurrences of unidirectional currents. On this basis, the velocity regime functions of the reciprocating currents are formed, from which the recurrences of the resulting longshore water transport are found. If necessary, the distribution of the resulting longshore current velocity along the studied shore section can be plotted with the needed recurrence.

In the next block of the scheme, the vertical distribution of the suspended sediment concentration is calculated. To calculate the profile of the suspended sediment concentration in the zone of unbroken and slightly deformed waves, the model suggested by Kos'yan and Pakhomov (1981) and described in Chapter 2 was used. The model considered in Section 6.5.2. was used for similar calculations in the surf zone. The major longshore transport of the suspended sediment occurs in the narrow nearshore zone between the depth of significant wave breaking (h_B) and the water edge. Thus, as a first approximation, the longshore discharge of suspended sediment on a unit of coast length can be expressed in terms of concentration and longshore current velocity values averaged over the cross-section of this zone (W):

$$Q_{ys} = \overline{VS}\Omega \qquad (7.14)$$

The value of Ω will be defined by the depth of significant wave breaking, which can be found in each particular case from $(H/h)_B = $ const. Thus, from the wave parameter regime functions, the mean concentration, longshore current velocity, and Ω can be found, from which Q_{ys} regime functions can be obtained.

7.3.2 Short description of the area

The calculation scheme has been successfully performed on the 120-km long section of Bulgarian Black Sea coast from Caliacra Cape to Emine Cape, which totals one-third of the overall sea border of Bulgaria. The well-known resort complexes famous for their excellent sand beaches are situated here; the development of the beaches is monitored constantly. Since 1971, the specialists from the Oceanology Institute of the Bulgarian Academy of Sciences (OI BAS) have measured the water depths regularly and sampled the bed sediment strictly along the range lines, down to the depth of 20 to 25 m, sometimes to 30 m.

Observational data served as a basis for the detailed bathymetric map of the region on a 1:2000 scale, with isobaths spaced one meter apart to the depth of 15 to 25 m. These maps, together with the spatial distribution of bed sediment, were amiably put at our disposal by the Institute of Oceanology Bulgarian Academia of Science.

Within the studied region, close to Caliacra, Balchic, Varna, and Obzor, four coastal hydrometeostations (CHMS) are located, at which four measurements of the wind direction and velocity were performed daily for a considerable period with subsequent interpolation of these data for the intermediate coastal sections. Observational data on wind velocities averaged over a period of more than 30 years and for eight compass points have been published in the Climate Reference Book (Vyatr, 1982).

The underwater slope of the studied area is rugged with submarine erosion-landslide terraces and bars. Rock outcrops in the form of erosion buttes or ridge-benches act as natural breakwaters. Erosion-landslide shores amount to 72% of the total length of the shoreline. The thickness of discontinuous sand bands amounts to 2 to 3 m. Accretional shores correspond to river mouths and constitute 28% of the total shoreline length. Land sliding and the erosion of shores, beaches, and the underwater slope are the main sources of benthic material. The bottom slope varies within the 0.01 to 0.03 range. The mean diameter of the bottom sediments varies from 0.188 to 0.750 mm.

For the purpose of predictive calculations, the studied region was divided into eight quasi-uniform sections (Fig. 113), the factors of division being: the relative straightness of the shoreline, its orientation, wave exposure directions, angle between isobath and shoreline, bottom gradients, and the uniformity of the underwater slope sediment.

7.3.3 Longshore discharge of suspended sediment and its regime functions

In accordance with the modular scheme described above, the velocities of the resulting longshore current, mean concentrations of the suspended sediment, cross-sections of longshore flow, and regime functions of these parameters have been calculated for each of the shore sections (Kos'yan and Phylippov, 1990).

The recurrence of "positive" (toward Caliacra Cape) or "negative" (towards Emine Cape) longshore discharges of suspended sediment is the summary recurrence of discharges along each of the relevant wave directions. On the basis of these data, the regime functions of "positive" and "negative" longshore discharges of suspended sediment were formed for each shore section.

The recurrence of each particular direction of summary longshore transport of suspended sediment was plotted next. The resulting discharge of the suspended sediment was calculated on the borders of each section as a sum of the sediment discharges of similar recurrence. By the plots of these regime functions, the resulting sediment transport with the given recurrence over the studied area can be estimated both qualitatively and quantitatively.

7.3.4 Prediction of the longshore discharge of suspended sediment

The results of the above calculations enable one to predict the longshore transport of suspended sediment on the studied section of Bulgarian coast at any time interval. The calculated longshore transport of suspended sediment with the recurrence $K_r = 24$ h/year on eight coastal sections is considered below as an example and compared with the observation data.

The distribution of suspended sediment at the studied area at the chosen recurrence was plotted from these data (Fig. 114).

The analysis of the distribution shows that two principal branches of the longshore flow of suspended sediment can exist. On sections I, II, and III the predominate transport is directed toward Caliacra Cape, while on sections IV,

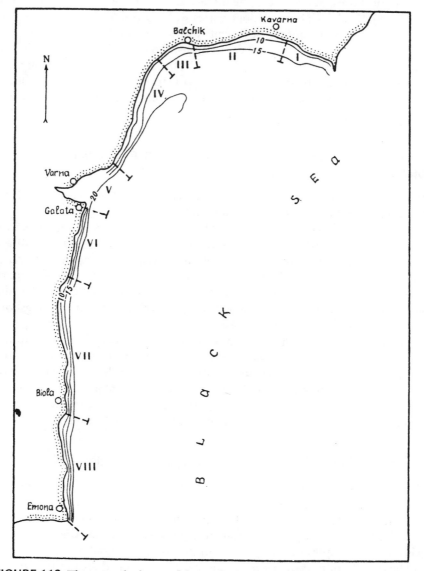

FIGURE 113 The general scheme of the studied sections of the coast.

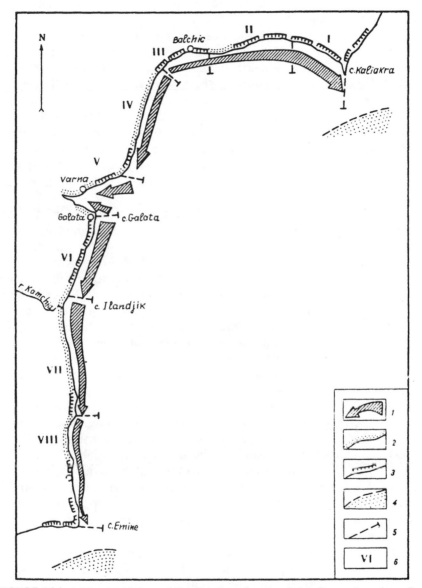

FIGURE 114 Arbitrary scheme of distribution of suspended sediment longshore flow at $k_r = 24$ h/y. 1) longshore sand flow; 2) accretional coast; 3) eroding coast; 4) accretional underwater body; 5) section boundaries; 6) section numbers.

VI, VII, and VIII, toward Emine Cape. On the fifth section, the sediment is transported along the Varna Bay shore and inward because of invading storm waves of east bearings.

Our calculations show that under certain storm conditions the suspended sediment flow along the shore of the first section is preferentially directed toward Caliacra Cape, its potential transport capacity amounting to 220 kg/s of sand. However, it is evident from the map of the bottom sediment that such amounts of sand are lacking in this area. As the coastal zone of the first section is characterized by an eroding shore without a beach and with a steep underwater slope with rock outcrops covered by bivalve colonies, it can be assumed to be the zone of sediment transit. The bed material brought from other areas or produced by shore and underwater slope erosion is transported toward Caliacra Cape and accumulated there. The sand shoal abeam with the Cape confirms this conclusion.

According to calculations, the longshore flow directed toward Caliacra Cape should exist on the second shore section, the potential discharge of the suspended sediment being 230 kg/s. Similar to the first section, the shoreline is slightly sinuous, with steep shores practically devoid of beaches over the entire length. Though there are some sands on the underwater slope, there is not enough material to saturate the longshore flow at the depths of 3 to 5 m even during storms. The sands appear at 5 to 7 m depths as narrow bands and lenses parallel to shore and migrating during severe storms due to the "significant" wave breaking. Eastward transport of these sands toward Caliacra Cape predominates. The transported sediments travel through the first section and are accumulated in the sand shoal abeam with Caliacra Cape at the depth of about 15 m.

The practical tasks of shore development can be solved effectively with the use of predictive estimates based on our calculations. Thus, it can be stated positively that the construction of artificial beaches on the first two sections is useless as the material will soon be outwashed and accumulated in another area. Being subjected to active erosion, the shore and the underwater slope are a source of bed material transported mainly toward Caliacra Cape, as our calculations have proved. Thus, the material cannot participate in longshore sediment transport toward Emine Cape, shown in the scheme given by Bulgarian researchers (Dachev and Cherneva, 1979).

The underwater slope of the third section is composed of sand sediment in the form of narrow longshore bands and lenses. The maximal thickness amounts to 2.5 to 3 m (Sorokin and Dimitrov, 1980). Narrow beaches appear in Balchic Bay, slightly cut into the steep shore. On the underwater slope, the sand sediment reaches the depth of 10 m. Storm waves beating this coast section generate the longshore flow directed toward Caliacra Cape, and the potential discharge of the suspended sediment can amount to 38 kg/s. If the input of sediment to this nearshore zone becomes lesser or stops currently, the existing beaches will evidently disappear. The time needed for this can be calculated from the regime functions of the resulting longshore transport of the suspended sediment.

From the known potential longshore discharge of the suspended sediment and the width of the main longshore flow, bottom relief and shoreline alterations can be predicted to provide for hydraulic works such as groins, breakwaters, jetties, marine channels, etc. Thus the construction of groins at the eastern end of Balchic Bay can result, on the one hand, in aggradation of the existing beach, and, on the other, in abrupt intensification of shore erosion behind the dike (on its eastern side facing Caliacra Cape).

Hydro- and morphodynamic conditions on the fourth coast section provide for longshore transport of the suspended sediment toward Emine Cape. The divergence zone on the border of the third and fourth sections is the arena of active erosion. Observational data showed that this zone corresponds to Cavaclar Cape and the northern end of the sand beach known as "The Golden Sands."

The beach and underwater slope sand is transported mainly toward Emine Cape, the potential discharge of the flow being about 180 kg/s of suspended sediment with $K_r = 24$ h/year. In the divergence zone, at the northern border of the section, minor flow of suspended sediment can exist simultaneously, being directed toward Caliacra Cape. The potential discharge of this flow is about 3.3 kg/s at the same recurrence. The berm of this section is composed of sand sediment. Two rivers, named Ekrine and Batowa, supply the beach and underwater slope with sand material. Between the water edge and the 7 to 10 m isobaths, large amounts of unsorted sands lie as longshore spots, lenses, and bands alternating in sediment coarseness. In the area of "The Golden Sands" a large rock shoal at depths of 5 to 15 m protects the beach from severe storms, acting as a natural breakwater.

Our calculations show that the decrease or complete termination of bed material input will gradually result in beach disappearance. If the present volume of sediment input holds, the reduction of beaches can only be caused by sediment attrition or drift into deep water, while with greater sediment input the beaches will naturally enlarge. The volume of material needed for beach enlargement can also be calculated from the regime functions of the resulting longshore discharge of the suspended sediment.

The shoreline of the fifth section delineates Varna Bay, subject mainly to eastern storm waves. Thus, the local longshore flow of suspended sediment can be formed here, being directed from east to west (i.e., inside the Bay). The potential discharge of the suspended sediment can amount to 98 kg/s. Within this section, longshore sandy bands alternate with the rock outcrops down to the depth of 15 m. The sand beach of this shore section has been formed and is now being fed by the sediment transported by the longshore flow from the fourth section. Estimated predictions suggest that the shoreline and the beach will remain intact at the present rate of the suspended sediment discharge. This conclusion is confirmed by the regime functions of the resulting discharge of the suspended sediment, which show that the suspended sediment is transported towards Varna Bay on the fourth section and from NE to SW on the fifth.

To preserve the beaches on the coast of Varna Bay, construction of hydraulic works (groins, piers, etc.) should be avoided, as they might cut off the existing longshore sediment flow feeding the beaches and result in their disappearance.

Longshore transport of the suspended sediment on the sixth section is formed by NE, E, SE, and S storm waves. The analysis of regime functions of the resulting longshore sediment transport reveals some consistent patterns. According to our calculations, the potential discharge of suspended sediment amounts to 340 kg/s and is directed toward Emine Cape. South of Galata Cape the shore is steep and hilly being locally cut by gullies and ephemeral streams supplying the sandy material for longshore transport. Abundant recent and ancient erosion terraces suggest that erosion is another source of sandy material. The underwater slope of this section is composed mainly by coarse sands while at the depth of 7 to 10 m the rock outcrops and landslide remnants are observed in the form of rocks, plates, and boulders partially covered by a thin layer of medium-grained sands.

Thus the shore of this section evidently acts as an active supplier of coarse material produced by erosion and slope processes. Being rather steep, the shore is closely approached by storm waves, which rework this material. It is transported southward by the powerful longshore flow. The peculiarities of longshore sediment transport should account for the rational development of this shore section. Thus, when predicting the coastal development, active erosion of the shore and underwater slope can be expected. If the volume of sandy material supplied by the available rivers and ephemeral streams, as well as by shore erosion, is known, practical measures aimed at shore stabilization could be recommended.

The seventh section is characterized by a generally straight, slightly rugged shoreline. The major part of the section is occupied by the sand beach, its width varying from 20 m in the north to 400 m in the center. The beach wedges out to the south gradually until it disappears completely at Black Cape. There are two rivers on this section, one of them, Kamchia, supplying 1.2 million tons of terrigeneous material annually. Unfortunately, the data on the present input of terrigeneous material are not available, but the observed beach reduction indicates a decrease of river sediment discharge.

According to our calculations, eastern storm waves can generate the longshore flow of suspended sediment directed toward Emine Cape with a potential discharge of 63 kg/s. This flow is evidently able to transport nearly half of Kamchia's annual sediment discharge. If the river sediment discharge decreases, the longshore flow would be fed by beach and underwater slope material, potentially resulting in beach erosion.

The eighth section is situated to the south of Black Cape and up to Emine Cape. According to our calculations, at $K_r = 24$ h/year, the longshore transport of suspended sediment can be directed toward Emine Cape, its potential discharge amounting to 40 kg/s. The sand material is concentrated on the underwater slope and in small beaches adjacent to the mouths of three small rivers. Active erosion of shores and ancient, now submerged, landslide blocks seems to be the dominant process on this section.

The sandy sediment transported alongshore is accumulated in the sand shoal abeam with Emine Cape. The analysis of regime functions of the resulting discharge of suspended sediment suggests that the sand material, once in shoal, cannot return back, but can participate in the longshore transport at Nesebrsky Bay coast.

The underwater slope and shoreline dynamics can be predicted by accounting for the amount of sandy material supply. If this amount is high enough to saturate the longshore flow, sandy material will be accumulated on the beaches existing in river mouths and on some capes (e.g., a narrow beach at Kochan Cape), or will build up new accumulative forms. Sediment deficiency will cause development of erosion and landward recession, which is actually happening just now on this section.

CHAPTER

8

Cross-Shore Sediment Transport and Variability of the Underwater Slope Profile

When the underwater slope is acted upon by the frontal storm waves approaching a uniform shoreline, the only sediment transport is normal to the isobaths. As a result of this transport, the underwater slope profile is changed in the zone of active wave effect. Cross-shore movement of bottom sediment is also possible with oblique wave approach to a shore with non-uniform bottom relief.

In contrast to the longshore sediment transport characterized by unidirectional movement at all levels above sea bottom, cross-shore movement of sand occurs both shoreward and seaward, the sign of the direction of motion reversing with depth at various points of the underwater slope. In order to evaluate the amount of transported sediment in this case, the profiles of concentration and of velocities in the zone of unbroken waves, breaking point, bore, and swash zone should be known. It has been shown in the previous chapters that the direct estimates of cross-shore discharge and simulation of the variations of the underwater slope profile in real conditions are still impossible. But along with predictions of the bottom slope deformation, the indirect methods for evaluating cross-shore sediment transport have been developed. These methods are briefly discussed below, together with the main approaches to bottom profile modeling and the results of natural observations on profile dynamics in storm conditions.

8.1 Sediment discharge

8.1.1 Experimental studies

It has been emphasized in Chapter 2 that all the forms of elementary hydrogenic processes are observed in various combinations when the underwater slope is affected by wave action. Due to the imperfection of the available measuring techniques, direct measurements of the sediment discharge normal to the shore presents a very difficult task. Special difficulties arise with the measurement of the net velocity profile that controls the direction of sand transport in depth at various points of the underwater slope. Attempts of such measurements with wave pressure sensors at the "Shkorpilovtzy" testing site showed that even during storms, the transporting velocities normal to shore did not exceed 1 to 3 m/s, measuring errors amounting to 100% and more at such low values. Laboratory measurements of the transporting velocity profile present similar difficulties, as it was noted in Chapter 1.

It has been stated above that the direct measurements of cross-shore discharge of sediment along the underwater slope profile are lacking, the only exclusion being the laboratory studies of bed sediment discharge above the flat bottom (Manohar, 1955; Kalkanis, 1964; Abou-Seida, 1965). In these studies the sediment discharge above the horizontal bottom was measured every half-period.

Measurements of sediment discharge above the sloping bottom deformed by wave action presents significant difficulties, and in this case the resulting discharge can be estimated by the method based on sequential measurements of the bottom profile. The main idea of this method is as follows: if the longshore conditions are constant (parallel isobaths, constant wave heights, absence of any longshore currents), then $\partial Q_y/\partial y = 0$, and profile variations, according to (7.1), will be defined by

$$\frac{\partial h}{\partial t} = \frac{1}{(1-n)} \frac{\partial Q_x}{\partial x} \tag{8.1}$$

where n is the bed sediment porosity. Integration of (8.1) with respect to x gives:

$$Q_x = Q_{x_0} + \int_{x_0}^{x} (1-n) \frac{\partial h}{\partial t} dx \tag{8.2}$$

Axis x is directed seaward with zero being at the shoreline. It is convenient to choose the point x_0 as a boundary point of the offshore profile, as the initial conditions for sand particle displacement are satisfied here and $Q_x = 0$ can be assumed. In this case (8.2) takes the form:

$$Q_x = \int_{x_0}^{x} (1-n) \frac{\partial h}{\partial t} dx \tag{8.3}$$

Thus, if the deformations $\partial h/\partial t$ are known, being determined by the successive measurements, the resulting sediment discharge normal to shore can be found for any point of the underwater slope. It follows from (8.3) that when the equilibrium profile is being built (usually defined as the ultimate stable profile constructed by infinite wave action), $\partial h/\partial t \to 0$ and $Q_x \to 0$.

This method is widely used in laboratory conditions as its basic requirements can be satisfied easily. Japanese researchers have been especially successful with its use (Watanabe, Rino, and Horikawa, 1981; Hattori and Kawamata, 1981; Shimizu, Saito, and Maruyama, 1985; Mimura, Ohtsuka, and Watanabe, 1986). Omitting the particular data obtained, we note only that sands of various coarseness were used in the experiments carried out in wave flumes varying from small-scale to large-scale (205 m long, 6 m deep, 3.4 m wide (Shimizu, Saito, and Maruyama, 1985)) in which waves of 2 m height could be generated.

Experimental laboratory studies showed that during profile modifications of an originally fine-grained sand bottom, the resulting seaward transport of suspended sand predominated. In the case of a coarse-grained bottom, the resulting shoreward transport was caused by the asymmetry of orbital velocities in the deformed waves. The results of Q_x measurement served as a basis for the empirical formulas discussed below.

Because of some uncertainties inherent in the modeling of sediment transport, direct extension of these results to natural conditions seems objectionable. It should be noted that in marine conditions, tidal and infragravity waves can play a significant role in transversal sediment transport during storms. Bed microforms also produce a notable effect as the suspension processes in their vicinity can control the resulting sand movement (Sunamura, 1981). Thus, the field studies of cross-shore sediment transport are needed to get a better understanding of its mechanism and evaluate it quantitatively. The main difficulty concerning the successive measurement technique lies in providing for uniform longshore conditions to make the use of equation (8.3) possible. The most thorough among these studies of this kind are those carried out by Hattori (1983) on the Pacific coast of Japan. The measurements were made on a site with roughly parallel isobaths, along the two range-lines normal to the shore, spaced 10 m apart. Metallic rods were rammed in the sea floor along the range lines with a 2 m interval, the entire distance amounting to 150 m. The seaward border of the range lines was at 2 m depth. The mean diameter of the bed sands was 0.18 mm. Water depths were measured at every rod hourly during a 24-hr period, each measurement requiring 15 min. Surface wave parameters were measured at seven points along the range line parallel to the depth measuring lines. The wave height of 0.35 m at the seaward end of the profile was roughly constant during the entire observation interval, the wave period being 7.3 s. Wave breaking through crest plunging was observed. The distribution of Q_x at various time instants is shown in Fig. 115.

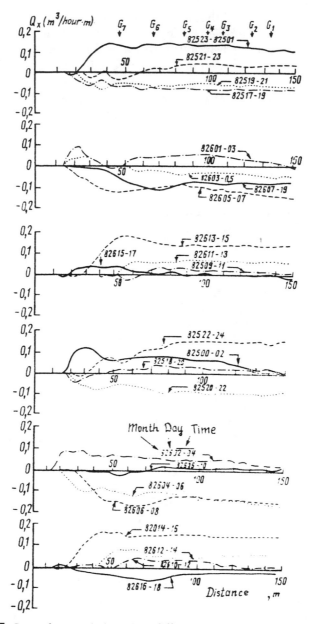

FIGURE 115 Cross-shore variations Q_x at different times (Hattori, 1983).

It is quite evident from the figure that under low surface waves the profile was far from equilibrium with dominating seaward sand transport. The direction of resulting sand movement was controlled mainly by the current direction in this zone. The distribution of Q_x along the range line in various instants of the tidal cycle is shown in Fig. 116. The resulting shoreward sediment transport corresponds to tides, while the seaward transport corresponds to ebbs. These results suggest that the tidal streams forming the field of transporting velocities, are the main factor controlling the direction of sand movement. This fact should be accounted for at sites with fine-grained sands easily suspended even by moderate waves.

In tideless seas the cross-shore direction of sediment transport can be controlled by infragravity waves; their possible role in the nearshore morphodynamics has been discussed in Chapter 2. Unfortunately, we do not know of any publications in which transversal sand transport by infragravity waves has been proved experimentally.

It should be noted as a conclusion that the evaluation of cross-shore sediment movement by the measurement technique has one uncertainty that should be accounted for when the data obtained by various authors are being compared or when the empirical relationships are being constructed. The fact is that as the bottom profile approximates an equilibrium state over time, the cross-shore sediment discharge under constant wave parameters in the seaward part of the profile tends to zero. Thus, in laboratory conditions, with an initial sloping flat bottom, the most intensive profile modifications and sediment transport occur in the very beginning of the experiment. As the profile approximates the equilibrium state, these processes decay. Thus, cross-shore

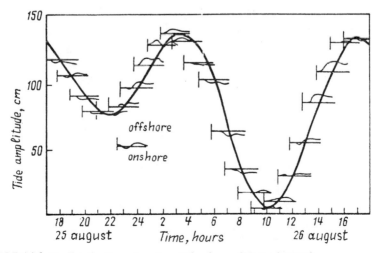

FIGURE 116 Q_x distribution at various tide phases (Hattori, 1983).

Q_x values appear to be time-variant under the same deep water wave parameters, and it is not clear which of them are most representative for the given waves. In real conditions, one never knows how far from equilibrium the given profile is, but this method can still be used to estimate the resulting sediment movement even if by an order of magnitude only.

8.1.2 Empirical formulas for evaluation of transversal sediment discharge

Among the known concepts for the quantitative evaluation of cross-shore sediment transport, those used for longshore transport prediction predominate. Available formulas, generally empirical in nature, are reviewed in detail in NDCP (1988). The main approaches to their construction are discussed below. Dimensional theory is most frequently used, according to which the dimensionless sediment discharge is a function of the Shields parameter. Madsen and Grant (1976) were among the first who used it to estimate the sediment discharge under wave conditions. They obtained the equation of bed sediment discharge averaged over a wave half-period using the laboratory data for a horizontal bottom obtained by Manohar (1955), Kalkanis (1964), and Abou-Seida (1965):

$$\frac{Q_x}{\overline{\omega_s d}} = 12.5 \Psi^3 \tag{8.4}$$

More recently this idea was used to obtain similar relationships for the resulting sediment discharge, Japanese scientists being the most consistent (Watanabe, Rino, and Horikawa, 1981; Sunamura, 1984; Yamashita, Sawamoto, and Yokoyama, 1984). In Watanabe et al. (1981) the resulting sediment discharge is assumed to be proportional to the difference between the Shields parameter and its value at the initiation of particle motion:

$$\frac{Q_x}{\overline{\omega_s d}} = A(\Psi - \Psi_{cr}) \tag{8.5}$$

The value of empirical coefficient A varies from 1 to 5 depending on the bed sediment composition, wave parameters, and the original bed slope, which is a disadvantage. Variability of the A value was avoided in later study (Watanabe, 1982). Additional analysis of the same data yielded another equation:

$$\frac{Q_x}{\overline{\omega_s d}} = 7(\Psi - \Psi_{cr})\Psi^{1/2} \tag{8.6}$$

In the above formulas, the Shields parameter is calculated by the amplitudes of near-bottom orbital velocity found from linear wave theory.

To analyze the data of his laboratory experiments, Sunamura (1984) used the factor of sediment mobility F_m and the Ursell parameter allowing for wave nonlinearity. The formula suggested for the resulting cross-shore sand transport is also valid for the splash zone and has the form:

$$\frac{Q_x}{\overline{\omega}_s \overline{d}} = -1.5 \times 10^{-7} U_r^{0.2} F_m (F_m - 0.13 U_r) \tag{8.7}$$

where $U_r = Hl^2/h^3$ is the Ursell parameter; the second term in brackets is an analog of the critical value of F_m.

The above empirical formulas have been obtained for monochromatic waves. Laboratory studies of irregular waves (Mimura, Ohtsuka, and Watanabe, 1986) showed that the distribution of experimental points in $Q_x/\overline{\omega}_s \overline{d}$ and Y coordinates varied insignificantly if Y was calculated from average or significant wave heights. Approximation of these data by (8.3) showed that under irregular waves, the value of empirical constant was lower than 7, varying from 1 to 5. This either suggests a lower effectiveness of sediment transport by irregular waves, or results from errors in determination of Q_x by the successive measurement technique (see the previous paragraph), with intervals between two successive measurements varying from one hour at the beginning to tens of hours in the end of each experiment.

To estimate the cross-shore sediment discharge and simulate the underwater slope profile, Bagnold's energy concept has been frequently used (Bowen, 1981; Bailard, 1981; Leont'ev, 1989, 1991). According to this approach, the instantaneous discharge of sediment is presented as a sum of near-bottom and suspended instantaneous discharge vectors (Bailard, 1981):

$$\vec{Q}_x(t) = \vec{Q}_{xB}(t) + \vec{Q}_{xs}(t) \tag{8.8}$$

$$\vec{Q}_{xB}(t) = \rho f_w \frac{C_B}{\tan \Phi} \left[\overline{|U(t)|^2 \vec{U}(t)} \right] - \frac{\tan \beta}{\tan \Phi} \overline{|\vec{U}(t)|^3} \vec{i} \tag{8.9}$$

$$\vec{Q}_{xs}(t) = \rho f_w \frac{c_s}{\omega_s} \left[\overline{|U(t)|^3 \vec{U}(t)} \right] - \frac{c_s}{\omega_s} \tan \beta \overline{|\vec{U}(t)|^3} \vec{i} \tag{8.10}$$

where c_B and c_s are coefficients of efficiency of bedload and suspended sediment transport, found empirically; $\vec{U}(t)$ = instantaneous nearbottom fluid velocity vector; $\vec{i}$ = unit vector directed upslope; F = friction angle; and $\tan \beta$ = the bottom slope.

The difference between the equations suggested by various authors results mainly from various estimates of the rule of gravity and the wave height distribution in the surf zone chosen for calculation of velocities. The underwater slope modeling is usually based on the energy approach, the principal aspects of which are discussed in the next paragraph.

8.2 Equilibrium profile modeling

Being of great practical significance, the prediction of underwater slope modifications during storms is one of the most important, though poorly studied problems of coastal engineering. The classical concept of a neutral line suggested by Cornaglia a century ago was later developed by a number of researchers (Longinov, 1963). According to this concept, the equilibrium of particles on the profile is controlled by the asymmetry of wave velocities directed shoreward, and by the down-slope displacement of particles by the force of gravity. Under long-term wave action upon the sloping bottom, the equilibrium profile is developed with the above factors balanced at each point and the resulting transport is equal to zero.

This concept was developed quantitatively by Bagnold (1963). On the basis of equality of sediment amounts transported by direct and reverse flows at the given point, the bottom slope was shown to depend on the ratio of energy lost during these phases. At equal losses (symmetric oscillations) the bottom slope tends to zero; the greater the asymmetry close to shore, the greater the bottom slope, the profile becoming concaved. These results hold true only for coarse sediment transported by traction. In case of a fine-sand bottom, wave transport of suspended sediment should be also considered. The models of equilibrium profile based on Bagnold's energy concept with accounting for both bottom and suspended transport, have been suggested by Bowen (1981), Bailard (1981), and Leont'ev (1989, 1991).

According to Bowen's model (1981), shoreward motion of sand particles is controlled by wave velocity asymmetry and transporting velocity in the near-bottom boundary layer, while their seaward movement is caused by the component of gravity tangential to the slope. The discharge of bottom and suspended sediment was taken in the form of (8.9) and (8.10) where $U(t)$ in the zone of unbroken waves was represented by the sum of oscillatory and transporting velocities. The latter was found in accord with the classical Longuet-Higgins equation, while the oscillatory component consisted of the first two harmonics. Calculations showed that if the sediments were transported in suspension only, the concaved bottom profile would form in equilibrium conditions ($Q_{xs} = 0$) with the depth $|x|^n$. Values of $n = 2/5$ correspond to the dominance of near-bottom wave transport, while $n = 2/3$ shows that the controlling factor is velocity asymmetry.

If the particles were transported solely as bedload, equilibrium conditions ($Q_{xB} = 0$) would be possible only on steep slopes typical of pebbly beaches.

With combined account of bedload and suspended sediment load ($Q_{xB} + Q_{xs} = 0$), the real slopes of a sand bottom were obtained at $U_m/\omega_s > 10$. Similar ideas were used by Bailard (1981). In his model, the equation for Q_{xs} was modified by a new interpretation of the efficiency coefficient c_s (in contrast with Bagnold's equation, in Bailard's formula this coefficient is divided by tan β). As a result of this modification, the influence of the gravity on suspension was reduced nearly by two orders of magnitude and the equilibrium state was reached at steeper slopes.

In the Leont'ev model (1988), like in the above ones, the wave approach is assumed to be normal to shore with uniform conditions along it. Waves are assumed to be steady, the velocity field being expressed as a sum of the first two harmonics and velocity determined by Euler's mass-transport velocity and the velocity generated by water level variations due to setup. His estimates showed that in the near-bottom boundary layer, these two components of the transporting velocity are in equilibrium and the transporting velocity tends to zero. Within the flow, water motion, and consequently that of suspended particles, is directed seaward. It follows from this model that the shoreward sediment transport in the near-bottom layer is caused by asymmetry of wave velocities, while the seaward suspension transport above the saltation layer results from the compensating reverse current. Averaging (8.8)-(8.10) over a wave period, Leont'ev defines the equilibrium profile ($Q_{xB} + Q_{xs} = 0$) with the equation:

$$\frac{dh}{dx} + r^{-1}\frac{8\bar{\omega}_s}{(H/h)^2(gh_0)^{1/2}}\left(\frac{h_0}{H}\right)^{1/2} - A_r = 0 \qquad (8.11)$$

where

$$A_r = \frac{16c_sU_m}{9\pi c_BU_{2m}}\tan\Phi \qquad (8.12)$$

where h_0 is the depth of wave breaking of 1% recurrence; U_{2m} is the second harmonic amplitude; r is the proportionality coefficient approximating a unit. In the surf zone $H/h = const = 0.8$ is adopted. Examples of equilibrium profiles calculated by this model for various coarseness of sediment are shown in Fig. 117. Points of inflection correspond to wave breaking points. The profile is generally S-shaped, concaved in the surf zone and convex seaward of the wave breaking zone. Slope steepness grows with sediment coarseness. The equilibrium profiles calculated by this model agree well with those observed in laboratory conditions at $A_r = 0.2$.

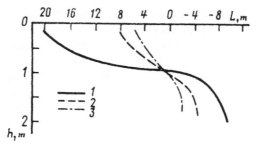

FIGURE 117 Calculated beach profiles at $h_0 = 1$ m, tan $\Phi = 0.5$, $A_r = 0.15$ (Leont'ev, 1988). $1 - \omega_s = 0.04$ m/s; $2 - \omega_s = 0.05$ m/s; $3 - \omega_s = 0.06$ m/s.

Equations (8.11) and (8.12) can be also used for irregular waves if $(H/h)^2$ values are averaged over a wave ensemble. Equilibrium profiles thus obtained are similar to those for regular waves, but in the latter the convex section is seaward of the wave breaking zone, while in the first it lies within the area of energy dissipation.

The basic principles of this model were used for profile prediction in the process of equilibrium profile development (Leont'ev, 1991). In this latter model, the bottom profile in various time instants is calculated by (8.1), the cross-shore discharge value Q_x being found from

$$Q_{xB} = \frac{9\pi c_B U_{2m}}{16 \tan \Phi U_m} D, \quad Q_{xs} = -\frac{c_s D}{\omega_s / U_n - \tan \beta} \qquad (8.13)$$

where $D = (4/3\pi) f_w p U_m^3$ is the rate of wave energy dissipation; $U_n = 1/8(H/h)^2(gh)^{1/2}$ is the mass-transport velocity; and U_m and U_{2m} are the amplitudes of the first and second harmonics of velocity. Boundary conditions are defined by the degree of difference between the real and equilibrium profiles, expressed through the slopes at the ends of the profile. Calculations by this model show that the bottom profile evolution is controlled by three parameters: the depth of wave breaking of 1% recurrence; sediment settling velocity ω_s; and the original profile slope. The first two control the variations in the inner part of the profile, while the slope of the original profile determines the deformation of its outer part. Temporal variations of the profile calculated theoretically are in good agreement with natural observations at the "Kamchia" testing sites (Nikolov and Pykhov, 1980).

The models in which the cross-shore sediment discharge is found from concentration and velocity profiles (Yang, 1981; Dally and Dean, 1984) seem to be most promising in terms of the underwater slope profile prediction. As in the above cases, the prediction is based on equation (8.1). In Yang's model

(Yang, 1981) the horizontal velocity field is presented as a sum of Fourier series for the oscillatory component and transporting velocity:

$$U(t) = C_F(1 + \bar{\eta}/h)\sum_{n=1}^{\infty} \beta_n \sin(n\omega t + \Psi_n) - U_n \qquad (8.14)$$

where U_n is the value of the mean velocity directed seaward; β_n and ψ_n are the amplitude coefficient and the phase angle, respectively, all the values being found empirically.

The suspension concentration is also presented as a Fourier series:

$$\bar{S}(z,t) = \bar{S}_x(z)\left[1 + \sum_{n=1}^{\infty} \beta_n \sin(n\omega t + \Psi_s)\right] \qquad (8.15)$$

The period-average suspension concentration is assumed to behave exponentially:

$$\bar{S}_x(z) = \bar{S}_c(x)\exp[-a_c\omega_s(z-c)/hU_{*m}] \qquad (8.16)$$

The value of the reference concentration $\bar{S}_c(x)$ depends on the point location on the profile and is controlled by the wave energy dissipation in the surf zone. Numerical calculations by this model give an S-shaped profile, and the agreement with laboratory data is achieved mainly by the fitting of empirical coefficients. This model, like those described above, does not give the bottom profile with an underwater bar.

In Dally and Dean's model (1984) the entire water column is divided into two layers. The height of the lower layer $\ell = \omega_s T$ is equal to the distance covered by a settling particle during the wave period, and the suspended sediment transport is controlled by transporting and orbital velocities. In this layer the particles are transported by orbital motion during wave crest transit after which they are supposed to resettle and remain on the bottom until the next crest comes. Particle motion during a wave period depends on the particle settling rate and the orbital velocity beneath the crest. In the upper layer, particles are also transported in suspension, their motion being controlled by the transporting velocity. The main driving force of the circulation in the plane normal to shore, which forms the structure of transporting velocity field, is the gradient of the transverse component of the radiational stress. The transporting velocity curve for a sloping bottom obtained on this basis has the form:

$$
\overline{U}_n(z) = \frac{gh}{8v} \frac{\partial h^2}{\partial x} \left[-\frac{3}{8}\left(\frac{z}{h}\right)^2 - \frac{1}{2}\left(\frac{z}{h}\right) - \frac{1}{8} \right] + \overline{U}_B \left[\frac{3}{2}\left(\frac{z}{h}\right)^2 - \frac{1}{2} \right]
$$
$$
- \frac{3}{2}\frac{Q_B^0}{h}\left[\left(\frac{z}{h}\right)^2 - 1 \right] \tag{8.17}
$$

where $\overline{U}_B$ is the flow velocity at the upper limit of the boundary layer; Q_B^0 is the resulting water flow between the bottom and the calm water level.

These parameters are calculated as:

$$
\overline{U}_B = \frac{U_0^2 \pi}{\lambda \omega}, \quad U_0^2 = \frac{H}{2}\left(\frac{g}{h}\right)^{1/2}, \quad Q_B^0 = \Psi_m\left(\frac{gHT}{h}\right) \tag{8.18}
$$

where H is the wave height at the given profile point, and Ψ_m is dimensionless value of the current line on the free surface, determined from the wave steepness at the chosen profile point H/ℓ_0 and the relative depth h/ℓ_0 (where ℓ_0 is the deep water wave length).

To describe the concentration profile, one of the classical solutions for progressive flow is used:

$$
\overline{S}(z) = \overline{S}_c \exp\left[-\frac{15\omega_s(z-c)}{h(\tau_B/\rho)^{1/2}} \right] \tag{8.19}
$$

The bottom shear stress is represented as a sum:

$$
(\tau_B/\rho)^{1/2} = \left[\frac{f_w H^2 g}{16h} \right]^{1/2} + \left[\frac{-2h}{\rho H(gh)^{1/2}} \frac{\partial(Ec_g)}{\partial x} \right]^{1/2} \tag{8.20}
$$

where the first term defines the orbital motion share and the second defines the effect of wave breaking. In (8.20) $\partial(Ec_g)/\partial x$ stands for the degree of energy dissipation due to wave breaking.

The calculations by this model, examples of which are shown in Fig. 118, give a qualitative representation of the typical profiles observed in natural and laboratory conditions. Unlike the above models, this one generates both normal S-shaped profiles and those with an underwater bar. Quantitative comparison with the laboratory measurements, however, leaves much to be desired. This is typical of all the models because of uncertainties inherent to the fitting of empirical coefficients determining the cross-shore

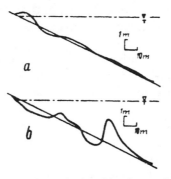

FIGURE 118 Typical beach profiles calculated by the model (Dally and Dean, 1984) $H_o = 1.4$ m; $H_B = 2.0$ m; $(H/h)_B = 1.3$; $T = 11.3$ s; a) $\omega_s = 0.0064$m/s, $\overline{d} = 0.4$mm; b) $\omega_s = 0.0023$m/s, $\overline{d} = 0.2$ mm.

sediment discharge. Despite this disadvantage, Dally and Dean's model seems to be the most promising for further improvement. Thus, it can be improved with the help of various approaches to modeling of undertow and concentration at the plunge point, developed recently and discussed in Chapters 1 and 2.

In the last decade, modern data and ideas about hydrodynamics and sediment transport have been used. A model reproduced an equilibrium beach profile in the surf zone with fairly good accuracy for time-averaged field data. A time-averaged zero-net sediment discharge for random waves is proposed. Outside the surf zone it does not show a realistic bottom slope.

Six models developed by European and Japanese scientists are based on nearly the same type of modular scheme for hydrodynamics and sediment transport but with different degrees of assumptions and empiricism. In that work it is shown that this concept of coastal profile modeling gives the pronounced bar formation under regular waves and the flattening of a steep profile under irregular waves.

The problem of cross-shore sediment transport has been developed intensively only in the last decade and is still far from solution. There exists a significant shortage of experimental data concerning cross-shore sediment transport. This problem is an essential task of future investigations. The most recent interesting data about low-frequency cross-shore transport on barred and non-barred beaches were obtained by Osborne and Greenwood (1992a,b). When modeling the equilibrium profiles, sediment transport by infragravity waves within the surf zone should be considered, especially for the profiles with underwater bars. When constructing numerical models, it is also advisable to allow for variations of sediment composition along the profile and during a storm, which will be shown to be typical for field conditions.

8.3 Field studies of the variability of bed sediment composition and underwater relief during a storm

Unfortunately, the correlation between grain-size variations and short-term profile deformations is poorly studied. Thus, the accurate quantitative prediction of underwater slope deformation under the known surface wave parameters is still impossible.

During the experiments "Shkorpilovtzy-85-88," great attention was paid to the studies of short-term deformations of the underwater slope profile accompanied by variations of bed sediment composition relative to the wave regime. The field observations enabled us to widen the existing ideas of the processes concerned and to obtain data on sediment transport along the bottom profile during a storm (Kos'yan, 1987; Nikolov and Kos'yan, 1990).

8.3.1 Short-term variations

The water depths were measured repeatedly along the pier at the "Shkorpilovtzy" testing site from the shoreline to a 5 m depth, measuring points being spaced 5 m apart. A long metallic rod graduated to 1 cm was used. A metallic plate welded to its lower end prevented the rod from biting deeper into the bottom sediment. The accuracy of measurements varied within 2 cm providing for reliable measurement of relatively small deformations of the bottom relief. The bed sediment was sampled concurrently with depth measurements. For this purpose, the surface sand layer 3 to 5 cm thick was scraped off the bottom with a cylindrical sampler 100 cm in diameter.

During the "Shkorpilovtzy-85" experiment, deformations of the underwater profile along the pier were measured from September 18 to October 23, 1985. Profile variations (Δh) were registered in cm relative to the bottom level on September 18, 1985, taken as a reference mark. Bottom sediment sampling began on September 25. A total of 16 measurements of the underwater profile were made and 17 series of bottom sediment were sampled. Dried samples of solid particles were subjected to 18-fraction grain-size analysis, the results of which were used for calculation of distribution parameters of sediment composition. The resulting data allow one to follow the temporal variations of these parameters at the same points of the underwater profile relative to the wave regime (Figs. 119 to 123).

Unfortunately, constant measurements of wave field parameters were not made, so that the time intervals could be characterized only by separate measurements of wave elements. In some intervals (at times amounting to several days) the wave regime was estimated visually.

From September 18–30, 1985, the weather was generally calm with slight waves being observed only on September 24, 1985. Slight waves lasted from

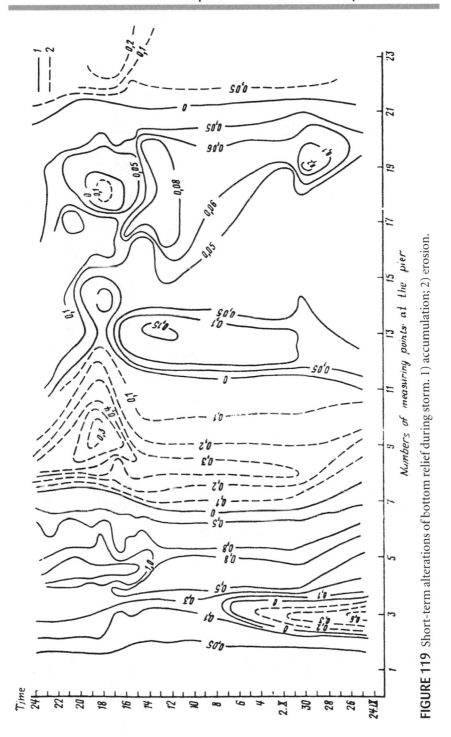

FIGURE 119 Short-term alterations of bottom relief during storm. 1) accumulation; 2) erosion.

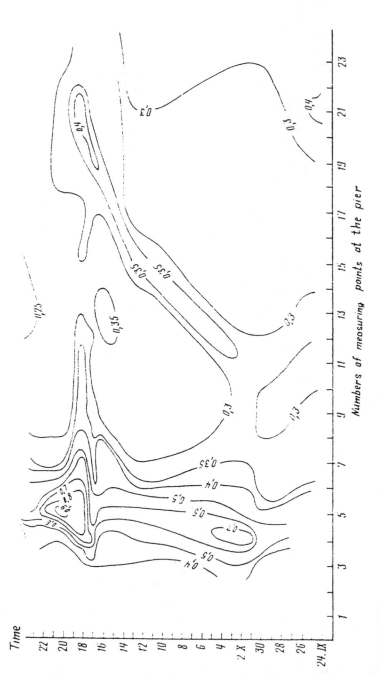

FIGURE 120 Temporal variations of mean grain size of bottom sediment.

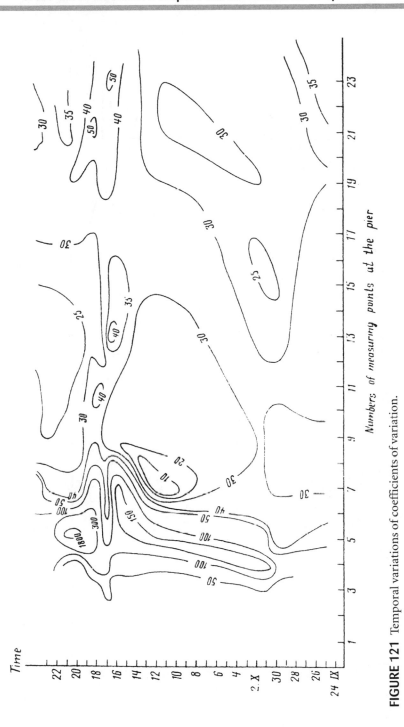

FIGURE 121 Temporal variations of coefficients of variation.

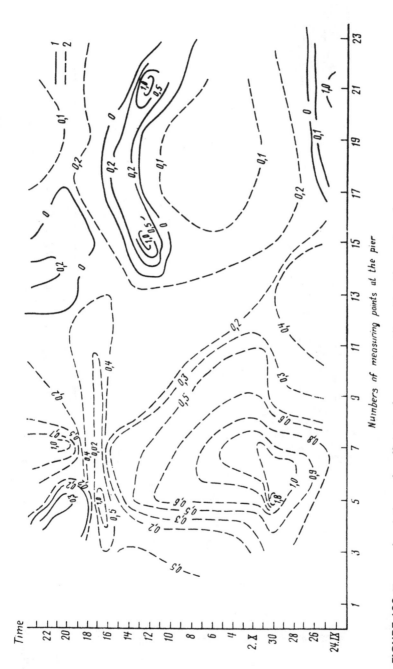

FIGURE 122 Temporal variations of coefficients of asymmetry. 1) positive; 2) negative values.

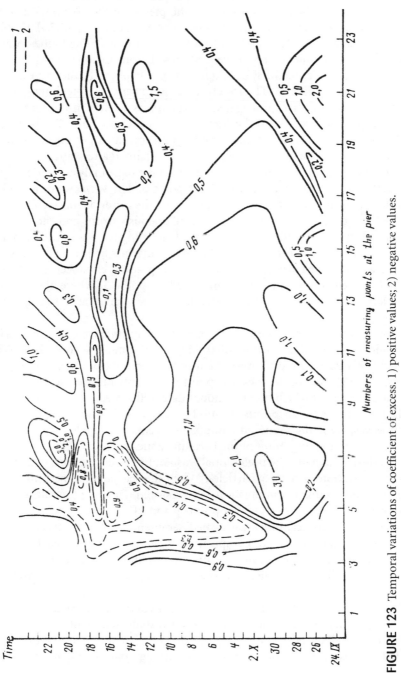

FIGURE 123 Temporal variations of coefficient of excess. 1) positive values; 2) negative values.

September 30 to October 11, 1985. In this observation period, waves were breaking nearly at the shore (point 5 of the pier). On October 11–12, 1985 larger wind waves were generally breaking between points 6 and 7. The wind direction changed on October 13–14 and the sea calmed. The highest waves of the entire experiment period were observed from October 15–18, 1985. Waves reached their maximal development on October 16, the largest ones breaking close to point 13. The rough waves were generally breaking close to points 7 and 8, and then propagated shoreward as a bore. In the remaining days of this period, nearly all the waves were breaking at point 7. The storm slackened on October 18 and intensified again for a short period on October 19. On October 20 the sea reduced to slight and calmed afterwards.

The following conclusion can be made from the data analysis. Under slight waves, the bottom aggraded gradually at point 5. The waves were plunging close to the shore where the bar was formed by the input of coarser material. The sediment grading was worsening simultaneously, from good to moderately good by Folk's (1974) classification. Under higher waves the breaking point migrated in the offshore direction resulting in further aggradation of the bar accompanied with particle fining and grading enhancement. When the sea calmed, the plunge point migrated to point 5, where the bottom aggradation recommenced, the sediments became coarser and their grading worsened.

Around point 7, the bottom was eroded slowly under slight waves with insignificant fining of bed sediment and better grading. The coarser material was evidently transported shoreward closer to the plunge point. Under higher waves, the migration of the breaking point to this area induced accumulation of coarser sediment which became moderately well sorted.

The bottom between points 9 and 14 was eroded insignificantly under slight waves, while the sediment composition remained stable. Under stronger storms and the migration of the breaking zone to points 7 and 8, bottom aggradation began with the preferential wash out of finer fractions and a relative concentration of coarser particles. When the sea calmed, the initial bottom level and sediment composition were restored. During observations, the sediment grading remained good at points 9 and 11.

At point 13, the accumulation of bed sediment continued under a calm and smooth sea, the sediment composition varying insignificantly. The composition characteristics generally remained constant throughout the observation period, while under higher surface waves the grading worsened insignificantly.

At point 15, the bottom level and sediment composition varied insignificantly. At points 17 to 21, the tendency toward slight sediment accumulation with a constant composition was observed. At higher waves, the bottom was eroded in this area, the bed sediment becoming coarser and more poorly graded.

At point 23, located at the seaward end of the trestle, bottom erosion was observed that intensified under higher waves with a preferential wash out of finer fractions and worsening of sediment grading.

Coefficients of asymmetry and excess were generally less than unity at all points indicating the small difference of the grain-size distribution curve from log normal. It should be noted that asymmetry coefficients were generally negative, while the excess coefficients were positive. This means that the maximum of the particle distribution density corresponds to fractions coarser than the mean size (remember that ϕ value grows with d value decrease). If compared with a normal distribution, the curve of sediment composition distribution has a higher and sharper peak.

Under slight waves the coefficients of asymmetry and excess decreased with distance from the breaking zone (seaward of point 7) and in the breaking zone itself, where the excess coefficient even took on negative values. Coefficients of asymmetry and excess were maximal in magnitude close to the plunge point.

Under higher waves, the breaking zone migrated seaward (points 7 to 9). The coefficients decreased in absolute value and remained nearly constant along the underwater slope profile seaward of the breaking zone, while shoreward of this zone (point 5) their absolute values grew rapidly. When the storm abated, the waves were breaking close to point 5 and in this case the coefficients of asymmetry and excess behaved similarly to those typical of slight waves.

The results obtained can be interpreted as follows: In the nearshore zone, the sediment differentiation is caused by turbulence generated at both borders of the flow, in the bottom boundary layer and on the surface during wave breaking. When sediment differentiation is caused by a single mechanism, the distribution of sediment composition is described well by a log-normal curve. When the mechanisms of differentiation are integrated, the sediment grading deteriorates rapidly, and its composition becomes heterogeneous, with a polymodal asymmetric distribution.

8.3.2 Bottom profile and sediment composition dynamics

During the experiment "Shkorpilovtzy-88" the studies of 1985 were continued and expanded. From October 11–24, a total of 23 measurements of the underwater profile were made and 19 series of bed sediment samples were collected. The approximate time of depth measurements and surface sediment sampling is shown in Fig. 124. In this figure and afterwards, the measuring points situated 5 m seaward of the points with the same numbers, are designated by index A.

During this period, the surface wave parameters were observed continuously but irregularly, depending on wave environment variations. When the

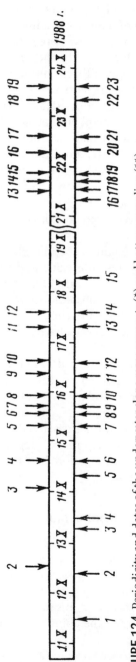

FIGURE 124 Periodicity and dates of the underwater slope measurements (*) and bottom sampling (**).

first measurements were made at 10 a.m. on October 13, the slight waves were breaking at points 7 and 5. By 1:30 p.m., the sea calmed and the breaking line migrated to the shoreline. Later this night, the northern wind and northeastern waves intensified. The width of the breaking zone amounted to 40 m, extending from point 11 to point 7. Most frequently the waves were breaking at point 9. On October 14 the sea calmed by 6:00 p.m., and by 8:00 p.m. the breaking zone migrated to points 7 to 9.

The next night (October 14–15), the swell intensified and separate waves were breaking at point 12, most of the breaking occurring at point 9. The next morning the wind intensified, and the breaking of large waves began at point 13. At 2:30 p.m. the breaking zone stretched further, the largest waves were breaking at point 15 and the small ones close to the shore line. At 10:30 p.m. the waves became more regular, the largest waves breaking at point 15 and sometimes even at point 17. The night of October 15 the wind and swell attenuated. By midday of October 16, the breaking of significant waves migrated to points 11 to 13, and to point 9 at 8.00 p.m., with only some of the waves breaking at point 11. By 10:00 p.m. the sea calmed a little, and the breaking line migrated shoreward. At 11:00 p.m. the waves were breaking at points 7 to 9. At night (October 16–17) the wind went down and all the waves were breaking close to shore. By 9:00 a.m. the sea was nearly calm. During daytime the swell intensified slightly and the breaking zone migrated to points 8 and 10, and by 5:00 p.m. of October 18 the sea became dead calm.

The second storm began in the morning of October 21. At midday the northern wind turned to the northeast, and by 8:00 p.m. the mass wave breaking occurred at points 7 to 8, while separate waves were breaking at point 9. The wind went down at night and the sea calmed, but at 10:30 a.m. next morning the swell intensified again, the wave breaking occurring mainly at point 11.

The last observation was made at 2:00 p.m. on October 23 when the majority of waves were breaking at point 9 and some of them at point 13.

While turning to interpretation of the underwater profiles obtained during this experiment, it should be noted that the measurements were generally made during the periods of maximal swell development. Maximal and minimal depths and d values registered at various points of the underwater profile during the observation period are shown in Figs. 125 and 126. The most significant fluctuations of these values were observed on the slope close to shore and around the underwater bar. Somewhat smaller variations were registered in the trough between the underwater bar and the near-edge slope, and seaward of this bar.

The examples of underwater slope profiles are shown in Fig. 127. Variations of the mean size of bed sediment along the profile at various times are shown in Fig. 128, and for some typical points during the entire observation period, in Fig. 129. The profiles from 1 to 8 inclusive were obtained during

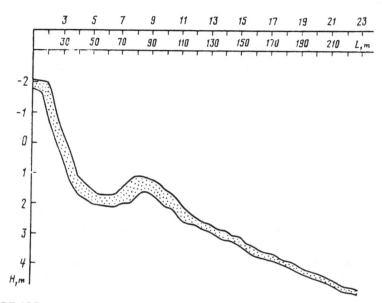

FIGURE 125 Maximal variations of the underwater profile during observation period (Shkorpilovtzy-88).

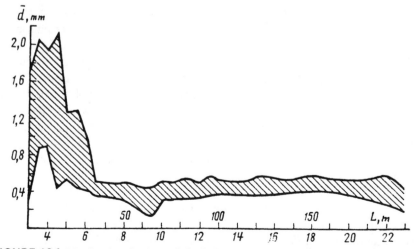

FIGURE 126 Maximal and minimal *d* values at various points of the underwater profile during the observation period (Shkorpilovtzy-88).

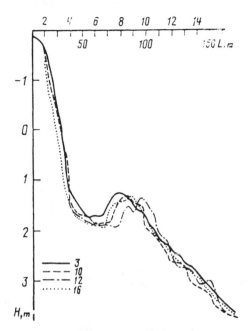

FIGURE 127 Examples of the underwater profile in various storm phases.

storm development. There are some variations which proceed successively as the swell intensifies. The shoreward side of the bar tends to aggrade (points 6 to 8), the water depth grows slightly at points 10 to 14, while the lower part of the near-edge slope subsides.

Due to further wave intensification, the largest waves begin to break above the bar top (point 7A), which gradually becomes the main breaking zone, resulting in destruction of the shoreward bar side (points 6A to 8). Then the reconstructed waves approach the shore and break again. Profile modifications are related to the growing thickness of the near-edge slope (points 3 to 4A), the outer side of the bar (point 8A to 10), and the upper part of the offshore slope (points 10 to 13). When the wave intensifies and the breaking zone migrates to the seaward slope (points 8 to 15), wave action creates shallow depths behind the breaking point. This part of the profile resembles a saw-like curve.

When the storm stabilized, the waves became nearly regular and were breaking at the same point (point 9) which resulted in significant erosion of the inner part of the bar.

More rapid attenuation than development of swell flattened the profile. The trough submerged and formed a nearly regular arc. By the end of the storm, the trough was filled in and the inner part of the bar aggraded due to the deposition of suspended particles and sediment displacement from the

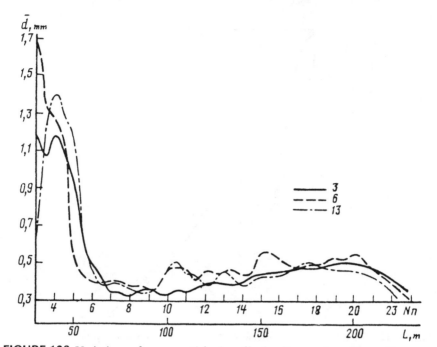

FIGURE 128 Variations of mean particle size of bed sediment along the profile at various time instants.

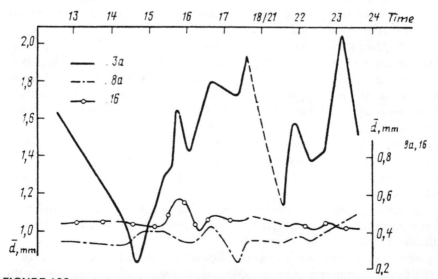

FIGURE 129 Variations of mean particle size of bed sediment in some typical points during the entire observation period.

bar top. The longer the phase of storm attenuation, the greater the trough filling and the closer to shore the bar is.

The analysis of profiles 16 to 18 obtained at the initial stage of the second storm, enables one to follow the profile variations after three days of calm sea. A slight sea with waves breaking close to shore reconstructed the profile nearly to its pre-storm form. After the storm, the relative height of the bar decreased and it became less steep, while the water depth at the seaward bar slope (points 11 to 23) increased. At the initial phase of the second storm, the waves acted upon the underwater slope only on its near-edge section, partially eroding it. With higher seas, the inner part of the bar was eroded (see Fig. 127).

Comparison of d distribution curves along the profiles of various series shows that the material coarseness changes abruptly close to points 5 to 6A. Shoreward of this line the sediment size d did not exceed 1.0 mm, while seaward of points 5 to 6A the d value was generally below 0.5 mm. This boundary was dynamic during the storm and tended to migrate shoreward during swell development and to return to its seaward position during its attenuation.

At the initial phase of swell development only the finest particles were displaced. When rising up the slope, they reduced the sediment coarseness close to shore (points 3 to 5) and on the bar (points 7 to 10).

When the compensating outflow develops, finer fractions near shore are carried offshore (series 4 and 5, points 3 and 4) and the mean grain size of the sediment increases. Further development of swell induces the suspension of finer sands and involves coarser fractions in cross-shore motion between points 7A and 15. At greater wave height, the same tendency is observed on the deeper section of the profile resulting in the growth of sediment coarseness at points 20 to 23 during maximal storm intensification. When the storm abates, the mean size of the sediment decreases on the seaward bar slope and in the trough due to the settlement of finer fractions from the water column. The coarser sediment is concentrated at the base of the slope close to shore. Sediment thickness at points 5 to 9 grows as a result of solid particle precipitation.

8.3.3 Some characteristics of cross-shore sediment transport

To consider the bottom relief variations during the storm, the underwater slope profile was divided into three zones, differing in the intensity of bottom deformation: the trough zone of the slope close to shore (zone I), the bar zone (zone II), and the seaward bar slope zone (zone III).

At relatively slight waves, surface waves are breaking only on the slope close to shore. Surf and compensating outflow thus formed, circulation is induced which transports finer fractions (if present in this zone) toward the trough, in which they settle or remain suspended. The slope being composed of coarse sand, the water depth does not change significantly at this stage.

As the storm develops and wave height increases, their deformation and breaking on the seaward bar slope generates more complex motion of water masses. In these conditions, finer particles can be transported seaward, beyond the bar top. The equilibrium profile is developed near the shore. For a short period sand particles are tracked shoreward by the bore. The compensating outflow acts for a longer period, during which finer particles cover a longer distance seaward and are accumulated there. The sediment is differentiated in a particular pattern along the slope. Sea calming causes settling of suspended particles on the near-edge slope, reducing the mean diameter of bed sediment.

The underwater bar zone is characterized by the high grading and uniformity of bed sediment. The insignificant growth of the d value was caused by erosion of the shoreward bar slope due to the transport of finer fractions on its seaward side. Alteration of bar height and its migration along the profile depend on preferential action of the compensating outflow upon the sediment of the shoreward bar side, on turbulence within the breaking wave suspending large amounts of sediment, and on the asymmetry of wave velocities. The predominance of the latter at the initial stage of swell development causes the shoreward migration of the bar due to the transport of bed sediment from its seaward side towards shore. Storm intensification results in wave breaking at the seaward bar slope, which is accompanied by the seaward migration of its top, so that sand particles are transported to its seaward side by a stronger compensating outflow. When the waves attenuate and the wave breaking zone migrates shoreward, this tendency is reversed. The absence of finer fractions around the bar can be explained by their removal by the compensating outflow.

The sediment particles behave similarly on the seaward bar slope and within the bar zone. The boundary between these zones can be traced by an abrupt drop of finer fraction percentage. Being a function of the bottom slope and wave length, the third zone width is dynamic. Under higher waves, this zone expands seaward. Bed sediment particles are preferentially transported up the slope and become coarser when reaching the bar. Under certain hydrodynamic conditions the particles of similar coarseness are being transported along the slope until they reach the equilibrium state.

The seaward part of the bar retains the slope profile when the storm abates abruptly, while the sediment coarseness everywhere tends to pre-storm values.

CHAPTER
9 | Conclusions

In this book we attempted to analyze the concepts of hydrogenous sediment transport by wave action in the nearshore zone, focusing primarily on modeling and natural studies. Following is a brief formation of the results of this analysis and some ideas concerning further studies.

The modeling of sediment transport is based on the simulation of hydrodynamic processes at various parts of the underwater slope, discussed in full detail in the first part of the book. The modeling of a turbulent boundary layer in a wave flow has advanced significantly due to perfect schemes for numerical modeling of turbulent currents. Correct specification of boundary conditions at the bottom remains the principal difficulty, which can be overcome with the help of careful laboratory studies of turbulent structure of the bottom boundary layer under various flow regimes.

During the last decade, the understanding of surf zone dynamics has advanced significantly. The models presented in this book enable the calculation, for practical purposes, of simple cases of nearshore circulation, longshore currents, and the vertical circulation. To solve the balance problem of sediment transport and bottom relief formation, three-dimensional models of circulation are needed, the first concepts of which have appeared recently. Within the framework of surf zone dynamics, great attention has been given lately to infragravity waves. Wide-scale field studies have revealed the main types of velocity fluctuations and their main parameters, though the generating mechanism is poorly studied. Modeling infragravity waves above the real bottom relief is still impossible and remains the main task of future research. Until this problem is solved, the prediction of underwater slope and beach relief is hardly possible since infragravity waves apparently play an important role in their formation.

Hydrodynamic processes in the wave breaking zone have also been studied intensively. The detailed laboratory research of Japanese scientists revealed the significant role of coherent vortex structures in the generation of small-scale turbulence and induction of wave outflow below a breaking wave. A number of promising models have been suggested for calculation of this out-flow current, while turbulence modeling still remains at the qualitative level, urgently calling for laboratory and in-situ measurements.

The state of knowledge of elementary sediment transport processes has been analyzed. Currently, only average characteristics of these processes can be calculated. The physical pattern of these processes remains vague, especially for time scales shorter than a wave period, while the turbulent structure of two-phase flows is still unknown. Recent experimental studies (Soulsby et al., 1984, 1986) indicate the great significance of the so-called "bursting" process of turbulence generation in the bottom boundary layer for understanding and evaluating the initial conditions of bed sediment displacement, particle suspension, and distribution within the water column. Bed sediment suspension by a translatory flow above a smooth bottom appears to be caused by the "ejection" of slow-moving liquid from the bottom into the water column, while liquid "inrush" toward the bottom determines the shearing stress responsible for the initiation of particle motion and bed sediment transport. Vortex structures defined by the "bursting" process of turbulence generation are supposed to be responsible for bed microform formation and transformation into three-dimensional forms. We know very little about the "bursting" process in a wave boundary layer, but the very first publications (Hino et al., 1983) indicated the presence of vortex coherent structures. Without a deep understanding of turbulence generation in a wave boundary layer, a thorough understanding and modeling of sediment transport is impossible.

The results presented in the first two chapters show that the parameters of elementary sediment transport processes and hydrodynamic fields can be predicted only in the simplest cases. In natural environments it is difficult to find conditions similar to those assumed in the highly simplified models. Therefore, natural observations play an important role in the understanding of real processes, and are vital to the process of refining these models. The studies of sediment transport carried out at various testing sites are discussed in detail. The testing sites are characterized briefly along with the methods used for experimental measurements.

The observed characteristics of suspended sediment distribution at time scales exceeding the wave period are discussed. Another important result of our studies is the detection of long-period variations of concentration at the bottom within the breaking zone caused by low-frequency waves. Our results show that sediment suspension at these exact periods determines the value of concentration averaged over longer periods. Research of the temporal variation of concentration and its relationship with long-period oscillations within the velocity field throughout the water column and at various sections of underwater profile can apparently lead to the modeling of suspended sediment motion in the coastal zone.

The review of studies dealing with mass sediment transport and transformations of the underwater slope has not revealed any new ideas on modeling these processes. Natural studies reveal the complex structure of the longshore discharge of suspended sediment, not allowed for in the existing models. From these models, only an order of magnitude for the discharge value can be deduced. Their further improvement will depend on the accuracy of modeling elementary coastal processes. Evaluative predictions of longshore sediment discharge for practical purposes are already possible, and examples are given in the book.

Further progress in the understanding of sediment transport processes will depend on the performance of natural experiments covering wide time intervals from tens of fractions of a second to several minutes and providing for synchronous velocity measurements. The solution to this task becomes feasible due to the development of optical and acoustic concentration and velocity meters, recently constructed and already being used for in-situ measurements.

References

Abou-Seida M.M. "Bedload Function Due to Wave Action." *Wave Res. Report HEL2-11*. University of California Hydraulic Engineering Lab, 1965.

Ackers P., White W.R. "Sediment Transport: New Approach and Analysis." *Journal of Hydraulic Div.* 11(1973): 2041-60.

Ahilan R.V., Sleath J.F.A. "Sediment Transport in Oscillatory Flow Over Flat Beds." *Journal of Hydraulic Engineering* 113(1987): 308-22.

Aibulatov N.A. *Study of Longshore Sand Sediment Transport in Sea*. Moscow: Nauka Publishing, 1966.

Aibulatov N.A. *Dynamics of Solid Substance in Shelf Zone*. L.: Gidrometeoizdat, 1990.

Aibulatov N.A., Kazakova V.K., Rudakova A.P. Quantitative Content and Composition of the Suspension on the *Mediterranean Northern African Shelf Geomorphology, Lithology and Lithodynamics*. Moscow: Nauka Publishing, 1982.

Aibulatov N.A., Kos'yan R.D., Antsyferov S.M. "New Things in the Sediment Flow Study in the Sea Coastal Zone." *Problems of Sea Shore Development*. Moscow: 10 USSR Academy of Science, 1989.

Aibulatov N.A., Kos'yan R.D., Orviku K.K. "Results of Lithodynamic Investigations from the Inhabited Laboratory 'Chernomor'." *Proceedings of Estonian Academy of Science, Chemistry, Geology*. 4(1974): 344-51.

Allen J.R.L. A Model for the Interpretation of Wave Ripple-Marks Using Their Wavelength, Textural Composition and Shape. *J. Geol. Soc.* 1979. Vol. 136 P.673-682.

Allen J.R.L. Field Measurements of Longshore Sediment Transport. *J. Coast. Res.* 1985. Vol.1, N 3. P.231-240.

Allen P.A. Reconstruction of Ancient Sea Conditions with An Example from the Swiss Molasse. *Mar.Geol.* 1984. Vol.60, N 1/4. P.1-16.

Altunin D.I. About Sediment Motion in Wave Flow. *Trans. of Moscow Hydro-reclaiming Inst.* 1975. Vol.42, P.18-22 (in Russian).

Antsyferov S.M. Experimental Research of Sediment Transport by Current: Thesis, M.: 1968. P. 115 (in Russian).

Antsyferov S.M. About the Distribution of Sediment Concentration in the Open Steady Flow. *Sediment Movement in Open Channels.* M.: Nauka Publ., 1969. P.191-196 (in Russian).

Antsyferov S.M. Methods of Determination of Suspended Sediment Concentration in the Upper Shelf. M.:P.P. Shirshov Institute, USSR , 1987. P. 64 (in Russian).

Antsyferov S.M., Basinski T., Kos'yan R.D. et al. Distribution of Suspended Sediment Over Coastal Zone of Lubiatowo. *Hydrodynamical and Lithodynamical Processes in the Marine Coastal Zone Inst. Hydroengin.* PAN, Warsaw; Poznan. 1980. P.119-127.

Antsyferov S.M., Basinski T., Kos'yan R.D., Onischenko E. Study of Suspended Sediment Distribution over the Coastal Slope. *Peculiarities and Transformation of Hydrodynamic Processes in the Nearshore Zone of the Nontidal Sea* ("Lubiatowo-74"), Gdynya, 1977. P.181-196 (in Russian).

Antsyferov S.M., Dachev V.D., Kos'yan R.D., Pykhov N.V. Suspended Sediment Distribution over the Coastal Slope of the Testing-Ground Kamchia. Interaction of Atmosphere, Hydrosphere, and Lithosphere in the Sea Nearshore Zone ("Kamchia-78"). S. Bulg. Ac. Sci. Publ., 1982. P.163-173 (in Russian).

Antsyferov S.M., Debolsky V.K. About the Distribution of Solid Particle Concentration and Size in the Open Flow. *Dynamics and Thermics of River Flows.* M.: Nauka Publ., 1973. P. 310-317 (in Russian).

Antsyferov S.M., Debolsky V.K. Some Peculiarities of Clastic Material Transport on the Shelf. *Lithodynamics, Lithology and Geomorphology of the Shelf.* M.: Nauka Publ., 1976. P.74-84 (in Russian).

Antsyferov S.M., Kos'yan R.D. Vertical Distribution of Sediment in the Steady Flow. *Meteorology and Hydrology.* 1976. N 8, P. 93-98 (in Russian).

Antsyferov S.M., Kos'yan R.D. Study of Suspended Clastic Material Movement in the Upper Part of the Shelf, behind the Bar Zone, *Oceanology.* 1977. Vol., N 3.P.497-505 (in Russian).

Antsyferov S.M., Kos'yan R.D. About the Coefficient of Eddy Diffusion of Sediment and About the Calculation of Distribution of Sediment Concentration in the Flow. *Meteorology and Hydrology,* 1979, N 5. P.72-79 (in Russian).

Antsyferov S.M., Kos'yan R.D. *Methodic Recommendations on the Study of Distribution of Suspended Sediment Concentration on the Upper Shelf.* M.: 1980. P. 29 (in Russian).

Antsyferov S.M., Kos'yan R.D. Sediment Suspended in a Stream Flow. *J. Hydraulic Div.* 1980. Vol. 106. P. 313-330.

Antsyferov S.M., Kos'yan R.D. Differentiation of Clastic Material in Sediment Carrying Flow. *Processes of Mechanical Differentiation of Clastic Material on the Upper Shelf.* M.: Nauka Publ., 1981.P.58-81 (in Russian).

Antsyferov S.M., Kos'yan R.D. *Suspended Sediment on the Upper Shelf.* M.: Nauka Publ., 1986. P. 224.

Antsyferov S.M., Kos'yan R.D. Dynamics of Suspended Sediment in the Nearshore Zone "Euromech-215." Mechanics of Sediment Transport in Fluvial and Marine Environment. Genoa, 1987. P.92-95.

Antsyferov S.M., Kos'yan R.D. Suspended Sediment Dynamics in Shallow Water. Proc. of Inter. Symp. on the Coastal Zone, with Special Reference to the Coastal Zone of China, Beijing, 1988. P.60-63.

Antsyferov S.M., Kos'yan R.D. Study of Suspended Sediment Distribution in the Coastal Zone. Coast. Eng. 1990 Vol. 14. P. 147-172.

Antsyferov S.M., Kos'yan R.D., Longinov V.V. Experimental Research of Sediment Shift Under the Action of Waves and Fair Stream. Transactions of Soyuznii Project. 1973. N 34 (40). P.108-115 (in Russian).

Antsyferov S.M., Kos'yan R.D., Onischenko E.L. To the Methods of Field Research of Suspended Clastic Material Motion Oceanology, 1975. Vol. 15, N 2, P.296-301 (in Russian).

Antsyferov S.M., Kos'yan R.D., Onischenko E.L., Pykhov N.V. About the Possibility of Suspended Sediment Concentration Measuring in the Sea by Means of Suspended Sediment Traps. *Oceanology.* 1977, Vol. 17, N. 6, P.1118-1122 (in Russian).

Antsyferov S.M., Kos'yan R.D., Onischenko E.L., Pykhov N.V. Substantiation of the Use of Suspended Sand Traps for the Measuring of Suspended Sediment Concentration and Composition. *Coastal Processes of the Nontidal Sea* ("Lubiatowo-76"), Gdansk, 1978. P.193-210 (in Russian).

Aono T., Hattori M. Experimental Study on the Spatial Characteristics of Turbulence Due To Breaking Waves. *Proc. 30th Jap. Conf. on Coast. Eng.* Tokyo, 1983. P. 25-29.

Aono T., Ohashi M., Hattori M. Experimental Study on the Turbulence Structure Under Breaking Waves. *Proc. 29th Jap. Conf. on Coast. Eng.* Tokyo, 1982. P. 159-163.

Asano T., Godo H., Iwagaki Y. Application of Low-Reynolds Number Turbulence Model to Oscillatory Bottom Boundary Layers. *Coast. Eng. Jap.* 1988. Vol. 30, N 2, P. 1-9.

Asano T., Iwagaki Y. Non-linear Effects on Velocity Fields in Turbulent Wave Boundary Layer. *Ibid.* 1986. Vol. 29. P.51-63.

Audin I., Shuto N. An Application of the K-E Model to Oscillatory Boundary Layers. *Coast. Eng. Jap.* 1988. Vol. 30, N 2. P.11-24.

Aukrust T., Brevik I. Turbulent Viscosity in Rough Oscillatory Flow. *J. Waterways, Port, Coast. and Ocean Eng.* 1985. Vol.111. N3 P.523-541.

Bagnold R.A. Motion of Waves in Shallow Water: Interactions Between Waves and Sand Bottom. *Proc. Roy. Soc. London A.* 1946. Vol. 187. P.1-16.

Bagnold R.A. Mechanics of Marine Sedimentation. *The Sea.* N.Y.: Inter Sci. Publ., 1963. Vol. 3. P. 507-528.

Bagnold R.A. An Approach to the Sediment Transport Problem from General Physics Geological Survey Professional Pap. 422-1. Wash. Govt. Print. Off., 1966. P. 72.

Bagnold R.A. The Nature of Saltation and Bed-Load Transport in Water. *Proc. Roy. Soc. London A.* 1973. Vol.332. P.473-504.

Bagnold R.A. Sediment Transport by Wind and Water. *Nordic Hydrology.* 1979. Vol. 10, N 5. P. 309-322.

Bailard J.A. On Energetics Total Load Sediment Transport Model for a Plane Sloping Beach. *J. Geophys. Res.* 1981. Vol. 86, N. 11, P.10938-10954.

Bailard J.A., Inman D.L. An Energetics Bedload Transport Model for a Plane Sloping Beach, Local Transport. 1981. *J. Geophys. Res.* Vol. 86, N. 3. P.2035-2043.

Bailey J.E. Particle Motion in Rapidly Oscillating Flows. *Chem. Eng. Sci.* 1974. Vol. 29. P.764-773.

Bakker W.T. Sand Concentration in an Oscillatory Flow. *Proc. 14th Int. Coast. Eng. Conf.* Copenhagen. 1974. P.1129-1148.

Barenblatt G.I. About the Suspended Particle Motion in the Turbulent Flow. *Applied Mathematics and Mechanics*. 1953. Vol. 17, N. 3.P.261-274 (in Russian).

Barenblatt G.I. About the Suspended Particle Motion in the Turbulent Flow which Occupies the Semi-Space or a Flat Channel with Finite Depth. *Applied Mathematics and Mechanics*. 1955. vol. 19, N. 1. P.61-88 (in Russian).

Bashkirov T. The Dynamics of the Sea Nearshore Zone. M.: *Sea Transport*, 1961. P. 220 (in Russian).

Basinski T. Measuring of Suspended Material Concentration with the Help of Gamma-globe and Suction-Type Device. *Interaction of Atmosphere, Hydrosphere and Lithosphere in the Sea Coastal Zone* ("Kamchia-77"). S.: Bulg. Ac. Sci. Publ. 1980a. P.277-284 (in Russian).

Basinski T. Objectives, Organization and Implementation of Lubiatowo-76 Program. Hydrodynamical and Lithodynamical Process in Marine Coastal Zone/Inst. Hydroengin. PAN. Warsaw, Poznan, 1980b. P.5-12.

Basinski T. *Field Studies on Sand Movement in the Coastal Zone*. Gdansk, 1989. P. 298.

Basinski T., Kasperovich Z., Onischenko E. Suction-Type Device for the Measuring of Suspended Sediment Concentration. *Coastal Processes of the Nontidal Sea* ("Lubiatowo-76"), Gdansk. 1978. P.279-290 (in Russian).

Basinski T., Levandowski A. Field Investigation of Suspended Sediment. *Proc. 14th Int. Coastal Engineering Conference*. Copenhagen. 1974. P.56-62.

Basinski T., Onischenko E., Pykhov N.V., Oktaba L. Radioisotopic Device for the Measuring of the Suspended Sediment Concentration Profile on the Shelf. *Coastal Processes of the Nontidal Sea* ("Lubiatowo-76"), Gdansk. 1978. P. 257-278 (in Russian).

Battjes A.J. Surf Similarity. *Proc. 14th Int. Coastal Engineering Conference*. Copenhagen. 1974. P.466-480.

Battjes A.J. Modeling of Turbulence in the Surf-Zone. *Proc. Symp. Model. Techn. ASCE*. 1975. N. 4. P.1050-1061.

Battjes A.J. Surf Zone Turbulence. *Proc. 20th Congress. JAHR. Semin. Hydrodyn. Coast. Zone*, Moscow, 1983. Vol.7. P.137-140.

Battjes A.J. Surf zone dynamics. *Ann. Rev. Fluid Mech*. 1988. Vol.20. P.257-297.

Bazilevich, N. Statistics of the Multi-Phase Systems. *Applied Mathematics*. 1969.V. 5. N 11. P. 365-371 (in Russian).

Beach R.A., Sternberg R.W. Suspended Sediment Transport in the Surf Zone: Response to Cross-Shore Infragravity Motion. *Mar. Geol*. 1988. Vol.80. P.61-79.

Belberov Z.K., Antsyferov S.M. Results of the Study of the Nearshore Zone Dynamics According To the Program of "Kamchia." M.:P.P. Shirshov Institute, USSR, Ac. Sci. 1985. P. 58 (in Russian).

Beloshapkov A.V. To the Problem of the Energy Method Development for the Calculation of Longshore Sediment Discharge. *Oceanology*. 1988a. Vol. 28, N 5. P.803-809 (in Russian).

Beloshapkov A.V. Comparative Analysis of Existing Methods and Shore Sediment Transport Calculation. Thesis. M. 1988b. P. 22 (in Russian).

Beloshapkova S.G., Beloshapkov A.V., Pykhov N.V., Antsyferov S.M. About the Initiation of Sediment Motion. *Oceanology*. S. 1990. N. 19. P.18-24 (in Russian).

Bhattacharya P.K. Sediment Suspension in Shoaling Waves: Ph.D. Thesis. Iowa, 1971. P.100.

Biesel F. Study of Wave Propagations in Water of Gradually Varying Depth. *Gravity Waves.* US Nat. But. St., Circ. 521. 1952. P.97-109.

Bijker E.W. Longshore Transport Computations. *J. Waterways Harbors and Coast. Eng. Div.* 1971. Vol. 97. P.298-309.

Bijker E.W. Some Considerations About Scales for Coastal Models with Movable Bed. Delft Hydr. Lab. 1967. N. 50. P. 142.

Bijker E., Kalwijk I., Pieters T. Mass Transport in Gravity Waves over the Sloping Beach. *Proc. 14th Int. Coastal Engineering Conference.* Copenhagen. 1974. P.447-465.

Biot M.A. Theory of Propagations of Clastic Waves in a Fluid-Saturated Porous Solid. *J. Acoust. Soc. Amer.* 1956. Vol. 28. P.168-191.

Birkemeier W.A., Dalrymple R.A. Nearshore Water Circulation Induced by Wind and Waves. *Proc. Symp. Model. Techn. ASCE.* 1975. N. 4. P.1062-1081.

Birkemeier W.A., Mason C. The Crab: A Unique Nearshore Surveying Vehicle. *J. Surv. Eng.* 1984. Vol. 110. N. 1. P.1-7.

Blondeaux P. Turbulent Boundary Layer at the Bottom of Gravity Waves. *J. Hydraulic Res.* 1987. Vol.25, N. 4. P.447-464.

Bosman J.J. Calibration of Optical Systems for Sediment Concentration Measurements. Delft. Hydraulic Lab. Rep. 1984. N. 716, pt. 5. P. 43.

Bowen A.J. Rip currents. Pt 1. Theoretical Investigations. *J. Geophys. Res.* 1969a. Vol. 74. P. 5467-5478.

Bowen A.J. The generation of Longshore Currents on a Plane Beach. *J. Mar. Res.* 1969b. Vol.27. P.206-215.

Bowen A.J. Simple Models of Nearshore Sedimentation: Beach Profiles and Longshore bars. *Coastline of Canada.* Halifax, 1981. P.1-11.

Bowen A.J., Guza R.T. Edge Waves and Surf Beats. *Ibid.* 1978. Vol. 83. P. 1913-1920.

Bowen A.J., Huntley D.A. Waves, Long Waves and Longshore Morphology. *Mar. Geol.* 1984. Vol. 60. P.1-13.

Bowen A.J., Inman D.L. Edge Waves and Crescenic Bars. *J. Geophys. Res.* 1971. Vol.76. P.8662-8671.

Bowen A.J., Inman D.L. Rip Currents, Pt 2. Laboratory and Field Observation. *Ibid.* 1969. Vol.74, N. 23. P. 5479-5490.

Bouden K.F. The Effect of Eddy Viscosity on Ocean Waves. *Phil. Mag.* 1950. Vol. 41, N 320. P.907-918.

Brebner A. Sand Bed-Form Length Under Oscillatory Motion. *Proc. 17th Coastal Engineering Conference.* Sydney, 1981. Vol. 2. P.1340-1343.

Brebner A., Resiedel P.H. A New Oscillating Water Tunnel. *J. Hydraulic Res.* 1973. Vol. 11,N 2. P. 107-121.

Brenninkmeyer B. In Situ Measurements of Rapidly Fuctuating, High Sediment Concentrations. *Mar. Geol.* 1976. Vol. 20, N. 2. P.117-128.

Brevik I. Oscillatory Rough Boundary Layers. *J. Waterways, Port, Coast. and Ocean. Div.* 1981. Vol.107, N. 3. P.175-188.

Brodkey R.S., Wallace J.M., Eckelmann H. Some Properties of Truncated Turbulence Signals in Bounded Shear Flows. *J. Fluid Mech.* 1974. Vol. 63. P.209.

Buchwald V.T., de Szoeke R.A. The Response of a Continental Shelf to Traveling Pressure Disturbances. *Austral. J. Mar. and Freshwater Res.* 1973. Vol.24. P.143-158.

BuevichY.A. About the Hydrodynamics of Homogeneous Suspensions. *Applied Mathematics and Technical Physics.* 1969. N. 6. P. 72-80 (in Russian).

Cantwell B.J. Organized Motion in the Turbulent Flow. *Annu. Rev. Fluid Mech.* 1981. Vol. 13. P.457-515.

Carlsen N.A. Measurement in the Turbulent Wave Boundary Layer Progr. Rept. Coast. Eng. Lab. and Hydraulic Lab. Techn. Denm. 1967. N. 14. P.2-3.

Carstens M.R. Accelerated Motion of Spherical Particles. Trans., *Amer. Geophys. Union.* 1952. Vol. 33, N 3, Pt. 1. P.713-721.

Carstens M.R., Neilson F.M. Evaluation of a Duned Bed Under Oscillatory Flow. *J. Geophys. Res.* 1967. Vol. 72. P.3053-3059.

Carstens M.R., Neilson F.M., Altinbilek H.D. Bed Form Generated in the Laboratory Under an Oscillatory Flow: Analytical and Experimental Study. US Army Corp. Techn. Memo. 1969. N. 28. P.1-39.

Chao B.T. Turbulent Behavior of Small Particles in Dilute Suspension. *Osterr. Ing. Arch.* 1954. Bd. 18, H 1/2. P. 7-21.

Clarke T.L., Lesht B.M., Young R.A. et al. Sediment Resuspension by Surface-Wave Action: an Examination of Possible Mechanisms. *Mar. Geol.* 1982. N 49. P.43-59.

Coakley J.R., Savile H.N., Pedrosa M., Lavocque M. Sled System for Profiling Suspended Littoral Drift. *Proc. 16th Coastal Engineering Conf.* Hamburg, 1978.

Coastal Sediments. New Orleans (La.), ASCE, 1987.

Coleman N.L. A Theoretical and Experimental Study of Drag and Lift Force on a Sphere Resting on a Hypothetical Stream Bed. *Proc. 12th Congress JAHR.* 1967. Vol. 3. P.185-192.

Coleman N.L. Flume Studies of the Sediment Transfer Coefficient. *Water Resour. Res.* 1970. Vol.6. N. 3. P.801-809.

Collins J.J. Inception of Turbulence at the Bed Under Periodic Gravity Waves. *J.Geophys. Res.* 1963. Vol. 68, N.21. P.6007-6014.

Coulter W.H. Pat. 2656508 U.S. Publ. 1953.

Crill S.I. Hydrodynamical Equations for Two-Phase Flow. *Proceedings of VNIIG.* 1969. V. 91 P. 72-83 (in Russian).

Crill S.I. Energy Equation for High Concentration Two-Phase Flow. *Hydromechanics. 1973.* N. 23. P. 71-77 (in Russian).

Dachev V., Kos'yan R.D., Pykhov N.V. Usage of Bathometers for Suspended Sediment Study in Sea. *Oceanology. S.* 1980, N. 6. P.69-79 (in Bulgarian).

Dachev V.D., Pykhov N.V. Vertical Distribution of Mean Size of Suspended Sediment in the Shoaling and Wave Breaking Zone. *Oceanology. S.* 1983. N. 11. P.56-60 (in Bulgarian).

Dachev V.D., Cherneva D.I. Estimation of Longshore Sediment Transport Rate Along Bulgarian Coast Between the Cape of Sivriburun and Burgas Bay. *Oceanology.* 1979. N. 4. P.30-42 (in Bulgarian).

Dally W.R., Dean R.G. Suspended Sediment Transport and Beach Profile Evolution. *J. Waterways, Port, Coast. and Ocean Eng.* 1984. Vol. 110. P.15-33.

Dally W.R., Dean R.G. Mass Flux and Undertow in a Surf Zone, by J.A. Swendsen-Discussion. *Coast. Eng.* 1986. Vol. 10. P.289-299.

Dalrymple R.A., Lozano C.J. Wave Current Interaction Models for Rip Currents. *J.Geophys. Res.* 1978. Vol. 83. P. 6063-6071.

Das M. Mechanics of Sediment Suspension due to Oscillatory Water Waves. Berkeley, Univ. Ca. 1971. Techn. Rep. N HEL-2-32.

Davies A.G. Field Observations of the Threshold of Sand Motion in a Transitional Wave Boundary Layer. *Coast. Eng.* 1980 Vol. 4. P.23-46.

Davies A.G. A Model of Oscillatory Rough Turbulent Boundary Flow. *Estuarine Coast. Shelf Sci.* 1986a. Vol. 23. N. 3. P.353-374.

Davies A.G. A Numerical Model of the Wave Boundary Layer. *Cont. Shelf Res.* 1986b. Vol.6, N. 6. P.715-739.

Davies A.G., Soulsby R.L., King H.L. A Numerical Model of the Combined Wave and Current Bottom Boundary Layer. *J.Geophys. Res.* 1988. Vol.93, N. Cl. P.491-508.

Davies A.G., Wilkinson R.H. Sediment Motion Caused by Surface Water Waves. *Proc. 16th Int. Coast. Eng. Conf.* Hamburg, 1976. P.1577-1595.

Dean R.G., Berek E.P., Gable C.G., Seymour S. Longshore Transport Determined by an Efficient Trap. *Proc. 19th Int. Coast. Eng. Conf.* Houston. 1984. P.954-968.

Deemter S.S., Laan E.T. Momentum and Energy-Balances for Dispersal Two-Phase Flow. *Appl. Sci. Res.* 1961. Vol.2, N. 10. P.108-109.

Deigaard R., Fredsøe J., Hadegaard B. Suspended Sediment in the Surf Zone. *J. Waterways, Port, Coast. and Ocean Eng.* 1986. Vol. 112. P.115-128.

Dement'ev M.A. General Equations and Dynamic Similarity of Sediment Carrying Flows. *Proc. VNIIG.* 1963. Vol.73 P.25-30 (in Russian).

Dement'ev M.A. About Hydraulic Calculation of Straight Line and Uniform Sediment Carrying Flows in the Current Nucleus. *Proc. VNIIG.* 1964. Vol.75. P.33-58 (in Russian).

Dement'ev M.A. Mechanisms of Ripple Formation on the Erodible Flow Bottom. *Proc. VNIIG.* 1982. Vol.154. P.69-73 (in Russian).

Dingler J.R. The Threshold of Grain Motion Oscillatory Flow in a Laboratory Wave Channel. *J. Sediment. Petrol.* 1979. Vol. 49. N. 1. P.287-294.

Dingler J.R. Wave-Formed Ripples in Nearshore Sands: Ph. D. Thesis. San Diego, 1974. P. 136.

Dingler J.R., Inman D.L. Wave Formed Ripples in Nearshore Sands. *Proc. 15th Int. Conf. Coast. Eng.* Honolulu. 1976. P. 2109-2126.

Dobroklonsky S.V., Mikhailova N.B. Filtration Flow Influence Upon the Intensity of Solid Particle Separation from the Bottom. *Hydrotechn. Engineering.* 1976. N 11.P.37-40 (in Russian).

Dou G. Sediment Transport and Bottom Stability in Water Flow. Thesis M. 1960. P. 254 (in Russian).

Downing J.P. Particle Counter for Sediment Transport Studies. *J. Hydraulic Div.* 1981. Vol.107. N. 11.

Downing J.P. Suspended Sand Transport on a Dissipative Beach. *Proc. 19th Int. Coast. Eng. Conf.* Houston. 1984. P.1765-1781.

Downing J.P., Sternberg R.W., Loster C.R.B. New Instrumentation for the Investigation of Sediment Suspension Process in the Shallow Marine Environment. *Mar. Geol.* 1981. Vol. 42. P.19-34.

Drake D.E., Cacchione D.A. Field Observations of Bed Shear Stress and Sediment Resuspension on Continental Shelves. *Cont. Shelf Res.* 1986. N.6. P.415-429.

Drake D.E., Cacchione D.A. Estimates of the Suspended Sediment Reference Concentration (Ca) and Resuspension Coefficient (γ) from Near-Bed Observations on the California Shelf. *Ibid.* 1989. Vol.9. P.51-64.

Dunin A.K. General Differencial Equations for Two-Phase Flow. *Proceedings of the Siberian Branch Academy of Science of Russia.* 1961. N. 10. P. 43-48 (in Russian).

Dupre W.R. Reconstruction of Paleo-Wave Conditions During the Late Pleistocene from Marine Terrace Deposits, Monterey Bay, California. *Mar. Geol.* 1984. Vol. 60, N. 1/4. P.435-454.

Du Toit C.C., Sleath J.F.A. Velocity Measurements Close to Ripple Bed in Oscillatory Flow. *J. Fluid Mech.* 1981. Vol.112. P. 71-96.

D'yachenko V.F. About One New Method of Numerous Solution of Non-Standard Problem with Two Space Variables. *J. of Calculating Mathem. and Mathem. Physics.* 1965. Vol.5., N. 4. P.680-688 (in Russian).

Dyer K.R. Velocity Profiles Over a Rippled Bed and the Threshold of Movement of Sand. *Estuarine Coast. Mar. Sci.* 1980. Vol.10. P.181-199.

Dyer K.R., Soulsby R.L. Sand Transport on the Continental Shelf. *Annu. Fluid Mech.*1988. Vol.20. P.295-324.

Dyhr-Neilsen M., Sorensen T. Sand Transport Phenomena on Coasts with Bars. *Proc. 12th Int. Coast. Eng. Conf.* 1970. P.855-866.

Dyunin A.K. General Differential Equations of Two-Phase Flow. *Proc. of Siberian Dept. USSR Ac. Sci.* 1961. N. 10. P.43-48 (in Russian).

Dzhrbashyan E.T. About Relative Velocity of Solid Particle Movement in Turbulent Flow. *Proc. of Arm. Ac. Sci.*, Ser. Techn. Sci. 1963 N .2/3. P.59-66 (in Russian).

Edge B.L., Magoon O.T., Baird W.F. "The Evolution of Breakwater Design." *Proceedings of International Conference on Waves.* ASCE. New Orleans. 1993.

Einstein H.A. The Bed-Load Function for Sediment Transport in Open Channel Flows. U.S. Dep. Agr. Soil Conserv. Serv. 1950. N. 1026.

Engelund F., Hansen E. *A Monograph on Sediment Transport in Alluvial Streams.* Copenhagen: Teknik forl., 1967.

Farber R. Investigations of Particle Motions in Turbulent Flow. Transport of Suspended Solids on Open Channels. Rotterdam: Balkema, 1986. P. 33-36.

Feedman B.A. About the Equations of Hydromechanics for Multicomponent Turbulent Medium. *Proc. Siberian Dept. USSR Ac. Sci.*, Ser. Techn. Sci. 1965. N. 2. P.133-135 (in Russian).

Feedman B.A., Lyatkher V.M. Turbulence Investigation by Filming Methods. *Dynamics and Thermics of River Flows.* Moscow. Nauka Publ. 1972 P. 109 126 (in Russian).

Fenton J.D., Abbott J.E. Initial Movement of Grains on a Stream Bed: The Effect of Relative Intrusion. *Proc. Roy. Soc. London A.* 1977. Vol. 352. P. 523-537.

Figurovsky N.P. *Sedimentographic Analysis.* M. Gostechizdat Publ. 1948. P. 332 (in Russian).

Fisher J.S., Pickral J., Odum W.E. Organic Detritus Particles. *Limnol. and Oceanogr.* 1984. Vol.24. P.529-531.

Folk R.L. Petrology of Sedimentary Rocks. Austin. 1974.

Frankl F.I. Energy Equation for Movement of Fluid with Suspended Sediment. *Rep. USSR Ac. Sci.* 1955a. Vol.102. N. 5. P.903-906 (in Russian).

Frankl F.I. Practice of Semiempiric Theory of Suspended Sediment Movement in Non-Uniform Flow. *Rep. USSR Ac. Sci.* 1955b. Vol.102, N. 6. P.1093-1096 (in Russian).

Frankl F.I. To the Theory of Suspended Sediment Movement. *Rep. USSR Ac. Sci.* 1953. Vol. 92, N. 2. P.247-250 (in Russian).

Fredsøe J. Turbulent Boundary Layers in Wave-Current Motion. *J.Hydraulic Eng.* 1984. Vol.110, N. 48. P.1103-1120.

Fredsøe J., Andersen O.H., Silberg S. Distribution of Suspended Sediment in Large Waves. *J. Waterways. Port, Coast. and Ocean Eng.* 1985. Vol.111. P.1041-1059.

Fredsøe J., Deigaard R. *Mechanics of Coastal Sediment Transport.* World Scientific. London. 1992. P. 367.

Friedlander S.K. Behavior of Suspended Particles in a Turbulent Fluid. *AIChE Journal.* 1957. Vol.3, N. 3. P.321-340.

Fukushima H., Mizogushi Y. Field Investigation of Suspended Littoral Drift. *Coast. Eng. Jap.* 1958. Vol.1. P.131-135.

Fukushima H., Mizogushi Y. A Study on the Sediment and its Measurement. *Ibid.* 1959. Vol. 2. P. 125-130.

Gable C.G. (Ed.). Report on Data from the Nearshore Sediment Transport Study. Experiment on Leadbetter Beach, Santa Barbara (Calif.) Jan./Feb. 1980. JRM Ref. 80-5, Univ. of Calif., Mar. Res. La Jolla (Calif.), 1981.

Gallagher B. Generation of Surf Beat by Non-Linear Wave Interaction. *J.Fluid Mech.* 1971. Vol. 49. P. 1-20.

Galloway J.S. Field Investigation of Suspended Sediment Clouds Under Plunging Breakers. *Estuarine, Coast. and Shelf Sci.* 1988. Vol.27. P. 119-130.

Gillie R.D. Canadian Coastal Sediment Study Final Report. Nat. Res. Council of Canada, Report Nc 232-17. 1985. P.171.

Gillie R.D. Sand and Gravel Deposits of the Coast and Inner Shelf East Coast, Northland Peninsula, New Zealand: Ph.D. Thesis. Canterbury, 1979.

Gizejewski Y., Dachev V., Pruszak Z. et al. Bed Microforms and Their Relation to Hydrodynamical Factors. *Rozpr. Hydrotech.* 1980. Vol.41. P.189-198.

Gizejewski E., Ronevich P., Roschenko V., Rudowski S. Microforms of the Black Sea Coastal Zone in the Area of Experimental Testing Ground. Interaction of Atmosphere, Hydrosphere and Lithosphere in the Coastal Zone ("Kamchia-78"). 1982. S. Bulg. Ac. Sci. Publ. P. 223-234 (in Russian).

Glenn S.M., Grant W.D. Suspended Sediment Stratification Correction for Combined Wave and Current Flows. *J. Geophys. Res.* 1987. Vol. 92. P.8244-8264.

Golikov V.I. Instrument for Measuring of Cloud Microstructure and Fog by Method of Narrow Angle. *Proc. of State Hydrol. Observatory.* 1964. N. 109. P.76-90 (in Russian).

Goncharov V.D. Dynamics of Channel Flows. L.:*Hydrometeoizdat*. 1962. P. 374 (in Russian).

Gordon C.M. Intermittent Momentum Transport in a Geophysical Boundary Layer. *Nature*. 1974. Vol.248. P.392-394.

Gordon C.M., Witting J. Turbulent Structure in Benthic Boundary Layer. Bottom Turbulence. Amsterdam.: Elsevier, 1977. P.59-82.

Graf W.H. Hydraulics of Sediment Transport. *Water Resour. Publ.* 1984. P. 513.

Granat N.L. Movement of Solid Body in Oscillating Flow of Viscous Fluid. *Proc. USSR Ac. Sci. OTN.* 1960. N. 1. P.70-78 (in Russian).

Grant W.D., Boyer L.F., Sanford L.P. The effect of Bioturbation on the Initiation of Motion Off Intertidal Sands. *J. Mar. Res.* 1982. Vol.40. P.659-677.

Grant W.D., Madsen O.S. Combined Wave and Current Interaction with a Rough Bottom. *J. Geophys. Res.* 1979. Vol.84. P.1797-1808.

Grant W.D., Madsen O.S. Movable Bed Roughness in Unsteady Oscillatory Flow. *Ibid.* 1982. Vol.87. P.469-481.

Grant W.D., Williams A.J., Gross T.F. A Description of the Bottom Boundary Layer at the HEBBLE Site: Low Frequency Forcing, Bottom Stresses and Temperature Structure. *Mar. Geol.* 1985. Vol.66. P.219-241.

Grass A.J. Initial Instability of Fine Bed Sand. *J. Hydraulic Div.* 1970. Vol.96. P.619-631.

Grass A.J. Transport of Fine Sand on a Flat Bed. Turbulence and Suspension Mechanics. *Euromech 48*. Denmark. 1974. P.33.

Greer M.N., Madsen O.S. Longshore Sediment Transport Data: A Review. *Proc. 16th Int. Coast. Eng. Conf.* Hamburg. 1978. P.1656-1674.

Grishin N.N. *Mechanics of Bottom Sediments.* M.: Nauka Publ. 1982. P. 160 (in Russian).

Guza R.T., Davies R.E. Excitation of Edge Waves by Incident on a Beach. *J. Geophys. Res.* 1974. Vol.79. P.1285-1291.

Guza R.T., Thornton E.B. Swash Oscillation on a Natural Beach. *J. Geoph. Res.* 1982. Vol.87. P.483-491.

Guza R.T., Thornton E.B., Holman R.A. Swash on Steep and Beaches. *Proc. 19th Int. Coast. Eng. Conf.* Houston. 1984. P.708-723.

Hagatun K., Eidsvik K.L. Oscillating Turbulent Boundary Layers with Suspended Sediment. *J. Geophys. Res.* 1986. Vol.91. N. 11. P. 13045-13055.

Hallermeier R.J. Sand Motion Initiation by Water Waves: Two Asymptotes. *J. Waterways, Port, Coast. and Ocean Div.* 1980. Vol.106. P.299-318.

Hallermeier R.J. Computation of Alongshore Sediment Transport. *J. Geophys. Res.* 1982. Vol.87. N. 8. P.5741-5751.

Hammond F.D., Heathershaw A.D., Longhorne D.N. A Comparison Between Shields Threshold Criterion in a tidal Channel. *Sedimentology.* 1984. Vol.31. P.51-62.

Hanes D.M., Huntley D.A. Continuous Measurements of Suspended Sand Concentration in a Wave Dominated Nearshore Environment. *Continent. Shelf Res.* 1986. Vol. 6, N 4. P.585-596.

Hanes D.M., Vincent C.E., Huntley D.A., Clarke T.L. Acoustic Measurements of Suspended Sand Concentration in the C2S2 Experiment at Stanhope Lane, Prince Edward Island. *Mar. Geol.* 1988. Vol. 81. P.185-196.

Hansen J.B., Swendsen I.A. A Theoretical and Experimental Study of Undertow. *Proc. 19th Int. Coast. Eng. Conf.* Houston. 1984. P. 2246-2262.

Haskind M.D. To the Sediment Theory About Heavy Particle Movement in Turbulent Flow. *Proc. USSR Ac.Sci. OTN.* 1956. N. 11 P.28-39 (in Russian).

Hasselmann K., Barnett T.P., Bouws E. et al. Measurements of Wind Wave Growth and Swell Decay During the Joint North Sea Wave Project (JONSWAP). Dt. Hydr. Ztschr. 1973. Suppl. A 8(12).

Hattori M. The Mechanics of Suspended Sediment due to Wave Action. *Coast. Eng. Jap.* 1969. Vol. 12. P. 111-132.

Hattori M. Field Study on Onshore-Offshore Sediment Transport. *Proc. 18th Int. Coast. Eng. Conf.* 1983. P. 923-940.

Hattori M., Aono T. Experimental Study on Turbulence Structure Under Spilling Breakers. *The Ocean Surface.* Dordrecht: Reidel, 1985. P. 419-424.

Hattori M., Kawamata R. Onshore-Offshore Transport and Beach Profile Changes. *Proc. 17th Int. Coast. Eng. Conf.* Sydney. 1981. P. 1175-1194.

Heathershaw A.D. "Bursting" Phenomena in the Sea. *Nature.* 1974. Vol. 248. P.394-395.

Heathershaw A.D. Measurements of Turbulence in the Irish Sea. *The Benthic Boundary Layer.* N.Y.: Plenum Press, 1976. P. 11-31.

Heathershaw A.D. Thorne P.D. Sea Bed Noises Revel Role of Turbulent Bursting Phenomenon in Sediment Transport by Tidal Currents. *Nature.* 1985. Vol. 316. P.339-342.

Hess F.R., Bedford K. Acoustic Backscatter System (ABSS): The Instrument and Some Preliminary Results *Mar. Geol.* 1985. Vol. 66. P. 357-379.

Hey A.E. On the Remote Acoustic Detection of Suspended Sediment at Long Wavelengths. *J. Geophys. Res.* 1983. Vol. 88. P.357-379.

Hill P.S., Nowell A.R.M., Jumars S.P. Flume Evaluation Relationship Between Suspended Sediment Concentration and Excess Boundary Sheer Stress. *J. Geophys. Res.* 1988. Vol. 93, N. C10. P. 12, 499-12, 509.

Hino M. Turbulent Flow with Suspended Particles. *J. Hydraulic Div.* 1969. Vol. 89, N. 4. P. 161-187.

Hino M. Theory on Formation of Rip Current and Caspidal Coast. *Proc. 14th Int. Coast. Eng. Conf.* Copenhagen. 1974. P. 901-919.

Hino M., Kashiwayanagi M., Nakayama A., Hara T. Experiments on the Turbulence Statistics and the Structure of Reciprocating Oscillatory Flow. *J. Fluid Mech.* 1983.Vol. 131. P.363-400.

Hinze I.O. Turbulence. M.: Physmatgiz Publ. 1963. P. 680 (in Russian).

Hjelmfelt A.T., Lenau C.W. Effect of Concentration on Sediment Distribution. *J.Hydraulic Div.*1969. Vol. 95, N. 5. P. 1775-1779.

Hjelmfelt A.T., Mockros L.F. Motion of Discrete Particles in a Turbulent Fluid. *Appl. Sci. Res.* 1966. Vol. 16, N. 2. P. 73-80.

Hoeg S., Schellenberger G. Messung der Wellenbedingten Sedimentan Fwireblung in the kustenaben Zone Acta hydrophys. 1971. Bd. 16, H. 2. P. 45-83.

Holman R.A. Infragravity Energy in the Surf Zone. *J.Geophys. Res.* 1981. Vol. 86. P. 6642-6650.

Holman R.A., Bowen A.J. Longshore Structure of Infragravity Wave Motions. *Ibid.* 1984. Vol. 89. P. 6446-6452.

Hom-ma M., Horikawa K. A Laboratory Study on Suspended Sediment due to Wave Action. *Proc. 10th Congr. JAHR. L.*, 1962. Vol. 1. P. 213-221.

Hom-ma M., Horikawa K. Suspended Sediment due to Wave Action. *Proc. 7th Int. Coast. Eng. Conf.* Berkeley, 1963. P.168-193.

Hom-ma M., Horikawa K., Kajima R. A Study of Suspended Sediment due to Wave Action. *Coast Eng.* Jap. 1965. Vol. 8. P. 85-103.

Hom-ma M., Horikawa K., Sonu C. A Study on Beach Erosion at the Sheltered Beaches of Katase and Kamakura. *Ibid.* 1960. Vol. 3. P. 45-61.

Horikawa K. *NearshoreDynamics and Coastal Processes.* University of Tokyo Press. Tokyo. 1988. P. 515.

Horikawa K., Watanabe A. Turbulence and Sediment Concentration due to Waves. *Proc. 12th Int. Coast. Eng. Conf.* Mexico. 1970. P. 479-482.

Horikawa K., Watanabe A., Katori S. Sediment Transport Under Sheet Flow Condition. *Proc. 18th Coast. Eng Conf.* Capetown, 1982. P. 1335-1352.

Hovem J.M. Attenuation of Sound in Marine Sediments Bottom Interacting Ocean Acoust. N.Y.: Plenum Press, 1980. P. 1-14.

Hsiao S.V., Shemdin O.H. Bottom Dissipation in Finite Depth Water Waves. *Proc. 16th Int. Coast. Eng. Conf.* Hamburg. 1978. P. 434-448.

Huntley D.A., Bowen A.J. Field Observation of Edge Waves. *Nature.* 1973. Vol. 243. P. 160-162.

Huntley D.A., Guza R.T., Thornton E.B. Field Observation of Surf Beat. 1. Progressive Edge Waves J.Geophys. Res. 1981. Vol.86. P. 6451-6466.

Huntley D.A., Kim C.S. Is Surf Bead Forced or Free? *Proc. 19th Int. Coast. Eng. Conf.* Houston. 1984. P. 871-885.

Hupfer P., Druet C., Kuznetsov O. International Experiment "Ekam 73" in the Coastal Zone of the Baltic Sea at Zingst (GDR) (in German) Beitrage zur Meereskunde. Leipzig.1974. N. 34. P. 61-64.

Ikeda S., Asaeda T. Sediment Suspension with Rippled Bed. *J.Hydraulic Eng.* 1983. Vol. 109. P.409-423.

Inman D.L. Wave Generated Ripples in Nearshore Sands. US Army Corp. Eng. Beach Eros. Board. Techn. Mem. 1957. N. 100. P.1-67.

Ismail N.M. Turbulent Transfer Mechanism and Suspended Sediment in Closed Channels. *Trans. Amer. Soc. Civ. Eng.* 1952. Vol.117, N. 409. P. 145-171.

Izumiya T., Horikawa K. Laboratory Study in Energy Dissipation due to Wave Breaking. *Proc. 29th Coast. Eng. Conf.* Japan. 1982. P. 115-154.

Jansen P.C.M. Laboratory Observation of the Kinematics in the Acreted Region of Breaking Waves. *Coast. Eng.* 1986. Vol. 9. P. 453-477.

Jobson H.E., Sayre W.W. An Experimental Investigation of the Vertical Mass Transfer of Suspended Sediment. *Proc. 10th Congr. IAHR.* Kyoto. 1969. Vol.2. P. 11-121.

Johns B. The Form of the Velocity Profile in a Turbulent Shear Wave Boundary Layer. *J. Geophys. Res.* 1975. Vol. 80, N. 36. P. 5109-5112.

Johns B. Residual Flow and Boundary Shear Stress in the Turbulent Bottom Boundary Layers Beneath Waves. *J. Phys. Oceanogr.* 1977. Vol. 7. P. 733-738.

Johnsson I.G. Measurements in the Turbulent Wave Boundary Layer. *Proc. 10th Congr. IAHR, L.,* 1963. Vol. 1. P. 85-92.

Johnsson I.G. On the Existence of Universal Velocity Distributions in an Oscillatory Turbulent Boundary Layer. Progr. Rept. Coast. Eng. Lab. Techn. Univ. Denm. 1966. N 12.

Johnsson I.G. Wave Boundary Layers and Friction Factors. Proc. 10th. Int. Conf. Coast. Eng. Tokyo, 1967. P. 127-148.

Johnsson I.G. A New Approach to Oscillatory Rough Turbulent Boundary Layer. Ocean Eng. 1980. Vol. 7. P. 109-152.

Johnsson I.G., Carlsen N.H. Experimental and Theoretical Investigations in an Oscillatory Rough Turbulent Boundary Layer. J.Hydraulic Res. 1976. Vol. 14. P.45-60.

Justesen P. Prediction of Turbulent Oscillatory Flow Over Rough Beds. Coast. Eng. 1988. Vol. 12. P.257-284.

Justesen P. Turbulent Wave Boundary Layers. Inst. Hydrodyn. and Hydraulic Eng. Techn. Univ. Denm. Spec. Pap. 1987. N. 43. P. 1-226.

Justesen P., Fredsøe J. Distribution of Turbulence and Suspended Sediment in the Wave Boundary Layers Progr. Rep. ISVA. 1985. N. 62. P. 61-67.

Kajiura K. A Model of the Bottom Boundary Layer in Water Waves. Bull. Earthquake Res. Inst. 1968. Vol. 46. P. 75-122.

Kalkanis G. Transport of Bed Material due to Wave Action. US Army Corps. Eng. CERC. Techn. Mem. 1964. N. 2. P.1-38.

Kalkanis G. A Model of the Bottom Boundary Layer in Water Waves. US Army Corp. Eng., Coast. Eng. Res. Center. Techn. Mem. 3. 1964. P. 68.

Kamphuis J.W. Friction Factor Under Oscillatory Waves. J. Waterways, Port, Coast. and Ocean. Div. 1975. Vol. 101, N. 2. P. 135-144.

Kamphuis J.W., Readshaw J.S. A Model Study of Longshore Sediment Transport Rates. Proc. 16th Int. Coast. Eng. Conf. Hamburg. 1978. P. 1656-1674.

Kana T.W. Surf Zone Measurements of Suspended Sediment. Proc. 16th Int. Coast. Eng. Conf. Hamburg. 1978. N 69. P. 1725-1743.

Kaneko A. The Wavelength of Oscillation Sand Ripples. Rep. Res. Inst. Appl. Mech. 1980. Vol. 28, N 88. P. 57-71.

Kaneko A. Oscillation Sand Ripples in Viscous Fluids. Proc. Jap. Soc. Eng. 1981. Vol. 307. P. 113-124.

Kaneko A., Honji H. Initiation of Ripple Marks Under Oscillating Water. Sedimentology. 1979. Vol. 26. P.101-113.

Kantarzhi I.G. et al. Recommendations on Computation of Wave Elements When Projecting Transport Hydrotechnical Construction. M.: TS IVIIS. 1983. P. 75 (in Russian).

Kennedy J.F. The Formation of Sediment Bedforms. Annu. Rev. Fluid Mech. 1969. Vol. 1. P. 147-168.

Kennedy J.F., Falcon M. Wave-Generated Sediment Ripples. MIT. Hydrodyn. Lab. Rept. 1965. N. 86. P. 55.

Kirlis V., Massel S., Zeidler R. Hydrodynamical Conditions in the Coastal Zone During "Lubiatowo-76." Proc. Int. Invest. "Lubiatowo-76." Hydrodynam. and Lithodynam Proc. in the Marine Coastal Zone. Warsaw-Poznan. 1980. P.37-48.

Kirlis V.I., Pavlyik K., Pustelnikov O.S. et al. Vertical Distribution of Suspended Sediment in the Near Water Edge Zone. Coastal Process of Nontidal Sea ("Lubiatowo-76"), Gdansk. 1978.N. 5. P.243-256 (in Russian).

Kitaigorodskiy S.A. About the Coefficient of Vertical Turbulent Exchange in Sea. *Proc. USSR Ac. Sci. Ser. Geophys.* 1957. N. 9. P.1118-1133 (in Russian).

Knoroz V.S. Noneroding Velocity for Noncohesive Soil and Factors that Determine it. *Proc. of VNIIG.* 1959, Vol.59. P.62-81 (in Russian).

Kolmogorov A.N. Equations of Turbulent Movement of Incompressible Liquid. *Proc. of USSR Ac.Sci. Ser. Phys.* 1942, Vol.6, N. 1/2. P.56-58 (in Russian).

Komar P.D. Beach Sand Transport Distribution and Total Drift. *J.Waterways, Port, Harbors and Coast. Eng.* 1977. Vol. 103. P. 225-239.

Komar P.D. Oscillatory Ripple Marks and the Evaluation of Ancient Wave Conditions and Environment. *J.Sediment. Petrol.* 1974. Vol. 44. P. 169-180.

Komar P.D., Inman D.L. Longshore Sand Transport on Beaches. *J.Geophys. Res.* 1970. Vol. 75, N. 30. P. 5914-5927.

Komar P.D., Miller M.C. The Initiation of Oscillatory Ripple Marks and the Development of Plane-Bed at High Shear Stresses Under Waves. *Ibid,* 1975. Vol. 45. P. 697-703.

Komar P.D., Miller M.C. The Threshold of Sediment Movement Under Oscillatory Water Waves. *J. Sediment. Petrol.* 1973. Vol. 43. P. 1101-1110.

Kos'yan R.D. About the Possibility of Application of Some Solution of the Problem About Suspended Sediment Movement. *Oceanology.* 1974, Vol.14, N 5. P.874-879 (in Russian).

Kos'yan R.D. Model of Vertical Distribution of Suspended Sediment Concentration in the Zone of Wave Energy Transformation. M.: 1983a. Reprint N. P.1055-83 (in Russian).

Kos'yan R.D. Theoretical Models of Sediment Carrying Flows. M.: 1983b. Reprint N 1025-83. P. 48 (in Russian).

Kos'yan R.D. Results of Lithodynamic Research in the Nearshore Zone on the North African Shelf. M. 1983c. Reprint. N. 5402-83. P. 64 (in Russian).

Kos'yan R.D. About the Calculation of Sediment Concentration Under Wave Action. *Water Resources.* 1984a. N 5. P.176-179 (in Russian).

Kos'yan R.D. Peculiarities of Calculation of Vertical Distribution of Suspended Sediment Concentration in Sea. *Oceanology* 1984b. Vol.24, N. 3. P.498-504 (in Russian).

Kos'yan R.D. Calculation of Mean Concentration of Suspended Sediment. Dynamics and Thermics of Rivers and Water Storage Basins. M.: Nauka Publ. 1984c. P.208-210 (in Russian).

Kos'yan R.D. Distribution of Suspended Sediment Concentration and Composition in the Plunging Point And in a Surf Zone. *Sofia. Oceanology.*1985a. N. 14. P.75-83 (in Russian).

Kos'yan R.D. Dependence of Passive Ripple Size on the Coarseness of Sediment Composing Those Ripple in the Coastal Zone. M.1986a.Reprint. N. 2465-B86. P. 16 (in Russian).

Kos'yan R.D. The Prognosis of Suspended Clastic Material Distribution on the Aquatory of the Region Under Study. *Oceanology.* 1986b. Vol.26, N. 6. P.981 (in Russian).

Kos'yan R.D. Vertical Distribution of Suspended Sediment Concentrations Seawards of the Breaking Zone. *Coast. Eng.* 1985b. Vol. 9. P. 171-187.

Kos'yan R.D. The Study of Wave Induced Sandy Bed Microforms in the Nearshore Zone. Mechanics of Sediment Transport in Fluvial and Marine Environments. Genova, 1987a. P. 41-44.

Kos'yan R.D. On the Dimensions of Passive Ripple Marks in the Nearshore Zone. *Mar. Geol.* 1988a. Vol. 80. P. 149-153.

Kos'yan R.D. Study of Sand Microforms in the Nearshore Zone. *Ibid.* 1988b. Vol. 83. P. 63-78.

Kos'yan R.D. Sand Bed Microforms in the Nearshore Zone During Storms. *Proc. Intern. Symp. on the Coastal Zone With Special Reference to the Coastal Zone of China.* Beijing, 1988c. P.17-19.

Kos'yan R.D. About Ripple Formation and Existence Under Wave Action in the Coastal Zone. *Water Resources.* 1987b. N. 1. P. 52-60 (in Russian).

Kos'yan R.D. Some Regularities of Sand Bottom Microform Distribution in the Coastal Zone. *Lithology and Minerals.* 1988d. N. 1. P.21-28 (in Russian).

Kos'yan R.D. Hydrogenous Transport of Sandy Sediments in the Coastal Zone of Nontidal Seas. D.Thesis. M. 1991. P. 467 (in Russian).

Kos'yan R.D. The Flows of Suspended Sandy Material Along the Section of Bulgarian Coast. Coastlines of the Black Sea. N.Y.: ASCE. 1993. P.488-501.

Kos'yan R.D., Antsyferov S.M., Efremov S.A. About the Applicability of Diffusion Theory to the Calculation of Sediment Distribution in Open Flow. *Meteorology and Hydrology.* 1976. N. 1. P.79-87 (in Russian).

Kos'yan R.D., Antsyferov S.M., Dachev V.D., Pykhov N.V. Calculation of Suspended Sediment Absolute Concentration Using Information Obtained by Suspended Sediment Traps. Interaction of Atmosphere, Hydrosphere and Lithosphere in the Coastal Zone ("Kamchia-78"). S. Bulg. Ac.Sci.Publ. 1982. P. 158-162 (in Russian).

Kos'yan R.D., Dachev V.D., Pykhov N.V. Some Peculiarities of the Concentration Field Formation of the Suspended Sediment in Breaking Zone. Interaction of Atmosphere, Hydrosphere and Lithosphere in the Coastal Zone ("Kamchia-77"). S.Bulg.Ac.Sci. Publ. 1980. P.266-276 (in Russian).

Kos'yan R.D., Kochergin A.D. About the Changeability of Concentration and Composition of Suspended Sediment by Shoaling and Breaking Waves. *Oceanology,* 1987. Vol.27. N. 2. P.286-295 (in Russian).

Kos'yan R.D., Pakhomov V.I. About Change of Suspended Sediment Concentration and Composition During Storm Period. *Oceanology,* 1979. Vol.19. N. 5. P.859-863 (in Russian).

Kos'yan R.D., Pakhomov V.I. One-Dimensional Distribution of Concentration of Sediment Lifted Above the Bottom by Waves. *Oceanology.* 1981. Vol.21. N. 5. P.865-871 (in Russian).

Kos'yan R.D., Pakhomov V.I., Pykhov N.V. About Vertical Suspended Sediment Distribution by Waves. *Oceanology.* 1978. V. 18 N. 5. P. 864-870 (in Russian).

Kos'yan R.D., Phylippov A.P. Methods of Prognosis of Suspended Longshore Sediment dDscharge. Problems and Methods of Investigation of Coastal Zone and Internal Basin Dynamics. Frunze. Ilim. 1987. P.14-15 (in Russian).

Kos'yan R.D., Phylippov A.P. Modular Circuit of Calculation of Suspended Longshore Sediment Transport. Rep. Conf. Results and Prospects of Physical-Geographic Investigations in Kirghizia. Frunze. 1988. P.10-11 (in Russian).

Kos'yan R.D., Phylippov A.P. Prognosis of Longshore Discharge of Suspended Solid Particles on the Bulgarian Coast Between Emine and Caliacra Capes. Practical

Ecology of Sea Regions. *The Black Sea. Kiev.*: Naukova Dumka Publ. 1990. P.177-182 (in Russian).

Kos'yan R.D., Pykhov N.V., Phylippov A.P. Vertical Distribution of Suspended Sediment Concentration and Composition in the Surf Zone. *Oceanology.* 1978. V. 18 N. 5. P. 1064-1069 (in Russian).

Kos'yan R.D., Onischenko E.L., Phylippov A.P. About Measuring of Variability of Bottom Sand Layer Thickness in Different Basins. *Meteorology and Hydrology.* 1988. N. 5. P.131-135 (in Russian).

Kotkov V.M. Experimental Research of Sand Ripples in Steady, Wave and Combined Flows. Complex Investigation of the World Ocean. M.: MGU Publ. 1979. Vol.2. P.86-90 (in Russian).

Kozlyaninov M.V. Hydrooptical Measuring in Sea. *Proc. of Inst. Oceanol.* USSR Ac.Sci. 1961, Vol.47. P.37-79 (in Russian).

Kraus N.C., Isobe M., Igarashi M., Sasaki T., Horikawa K. Field Experiment on Longshore Sand Transport in the Surf Zone. *Proc. of 18th Coast.Eng. Conf.*, ASCE. 1982. P.969-988.

Kraus N.C., Sasaki T.O. Effect of Wave Angle and Lateral Mixing on the Longshore Current. *Coast. Eng. Jap.* 1979. Vol. 22. P. 59-74.

Kril S.I. Equation of Energy for Highly Concentrated Sediment Carrying Flow. *Hydromechanics.* 1973. N. 23. P.71-77 (in Russian).

Kril S.I. Equation of Hydrodynamics for Two-Phase Mixtures. *Proc. VNIIG.* 1969 Vol. 91. P.72-83 (in Russian).

Krumbein W.C. Size Frequency Distribution of Sediments. *J.Sediment. Petrol.* 1934. Vol. 4. P. 65-77.

Kuchanov S.I., Levich V.G. Energy Dissipation in the Turbulent Gas Which Contains Suspended Particles. Rep. USSR Ac.Sci. 1967.Vol.174, N. 5. P.1033-1036 (in Russian).

Kuznetsov S.Y., Speransky N.S. Dynamic Characteristics of VDK-Sensor as a Velocity Meter. *Oceanology.* 1986. Vol.26, N. 2. P.335-341 (in Russian).

Lamb H. *Hydrodynamics.* Cambridge Univ. Press, 1932. N. 15. P. 738.

Larsen L.H., Sternberg R.W., Shi N.E. et al. Field Investigations of the Threshold of Grain Motion by Ocean Waves and Currents. *Mar. Geol.* 1981. Vol. 42. P. 105-132.

Lavelle I.W., Mofjield H.O. Effects of Time-Varying Viscosity on Oscillatory Turbulent Channel Flow. *J.Geophys. Res.* 1983. Vol. 88, N. 12. P. 17607-17616.

Lees B.J. Relationship Between Eddy Viscosity of Seawater and Eddy Diffusivity of Suspended Particles. *Geo-Mar.Lett.* 1981. Vol. 1. P.249-254.

Leont'ev I.O. Review of Modern Notions About Water Circulation Caused by Waves in the Coastal Zone. Lithodynamics and Hydrodynamics of Ocean Contact Zone. M.: Nauka Publ. 1981. P.128-153 (in Russian).

Leont'ev I.O. Longshore Sediment Transport Under the Irregular Waves Near the Shoal Coast. *Oceanology,* 1985. Vol.25, N. 4. P.638-644 (in Russian).

Leont'ev I.O. About Horizontal Nearshore Circulation Under Irregular Waves. *Water Resources.* 1987. N. 5. P.16-22 (in Russian).

Leont'ev I.O. *Dynamics of Surf Zone.* M. P.P. Shirshov Institute USSR Ac.Sci. 1988. P.184 (in Russian).

Leont'ev I.O. Randomly Breaking Waves and Surf Zone Dynamics. *Coast. Eng.* 1988. Vol. 12. P. 83-103.

Leont'ev I.O. About Possibility of Prognosis of Submerged Coastal Slope Profile Deformation. *Oceanology,* 1991. Vol.31, N. 5. P.735-743 (in Russian).

Leont'ev I.O., Efremov S.A.,Speransky N.S., Stoyanov L.D. Some Features of Distribution of Mass-Transport Velocities on the Testing-Ground of "Kamchia." Interaction of Atmosphere, Hydrosphere and Lithosphere of the Coastal Zone ("Kamchia-78"). S. Bulg. Ac.Sci. Publ. 1982. P.112-118 (in Russian).

Leont'ev I.O., Pykhov N.V. Investigation of Longshore Sediment Transport in the Coastal Zone. *Water Resources.* 1988. N. 4. P.53-63 (in Russian).

Leont'ev I.O., Speransky N.S. Investigation of Longshore Water Transport in the Coastal Zone. *Oceanology,* 1979a. Vol.19, N. 4. P.686-691 (in Russian).

Leont'ev I.O., Speransky N.S. Water Transport in the Nearbottom Layer of the Wave Flow in the Coastal Zone. *Oceanology.* 1979b. Vol.19. N. 5. P.854-858 (in Russian).

Leont'ev I.O., Speransky N.S. Study of Compensated Counter-Flow in the Coastal Zone. *Water Resources.* 1980. N. 3. P.122-131 (in Russian).

Lesht B.M., Clarke T.L., Young R.A. et al. An Empirical Relationship Between the Concentration of Resuspended Sediment and Near Bottom Wave Orbital Velocity. *Geophys. Res. Lett.* 1980. Vol. 7. P. 1049-1052.

Levich V.G., Myasnikov V.P. Kinematic Model for Boiling Layer. *Applied Mathem. and Mechanics.* 1966. Vol.30, N. 3. P.467-475 (in Russian).

Levin B.M. Mikhailova N.A. Orlov Bathometer Application for Measuring of Turbidity in Laboratory. *Proc. Moscow Inst. of Rail Transport Engineer.* 1960 N. 107. P.47-69 (in Russian).

Liu P.L.F, Dalrymple R.A. Bottom Frictional Stress and Longshore Currents due to Wave with Large Angles of Incidence. *J. Mar. Res.* 1978. Vol. 36. P. 357-375.

Liu V. Turbulent Dispersion of Dynamic Particles. *J. Meteorol.* 1956. Vol. 13, N. 4. P. 399-405.

Lofquist K.E.B. Measurements of Oscillatory Drag on Sand Ripples. *Proc. 17th Int. Coast. Eng. Conf.* Sydney, 1980. P. 3087-3106.

Lofquist K.E.B. Sand Ripple Growth in an Oscillatory Flow Tunnel. US Annual Corp. Eng. CERC. Techn. Pap. 1978. N. 78-5. P. 101.

Longinov V.V. Dynamics of the Coastal Zone of the Nontidal Seas. M.: USSR Ac. Sci. 1963. P. 379 (in Russian).

Longinov V.V. Review of Calculation Methods for Longshore Sediment Transport in the Coastal Zone. *Proc. SojuzMorNIIProject.* 1966. N. 14 (20). P.40-81 (in Russian).

Longinov V.V. Determination of Suspended Sand Concentration by Photoelectric Method in the Coastal Zone Under Wave Influence. *Proc. SojuzMorNIIProject.* 1968. N. 20 (26). P.82-92 (in Russian).

Longinov V.V., Pykhov N.V. Lithodynamic Ocean Systems Lithodynamics and Hydrodynamics of the Ocean Contact Zone. M.: Nauka Publ. 1981. P.3-64 (in Russian).

Longuett-Higgins M.S. Longshore Currents Generated by Obliquely Incident Sea Waves. *J. Geophys. Res.* 1970. Vol. 75 (a, b), Pts. 1 and 2. P. 6778-6801.

Longuett-Higgins M.S. Mass Transport in Water Waves. *Philos. Trans. Roy. Soc. London A.* 1953. Vol. 245. P. 535-581.

Longuett-Higgins M.S. The Mechanics of Surf Zone. *Proc. 13th Intern. Conf. Theor and Appl. Mech.* Moscow, 1973. P. 212-228.

Longuett-Higgins M.S., Stewart R.W. Changes in the Form of Short Gravity Waves on Long Waves and Tidal Currents *J. Fluid. Mech.* 1960. Vol.8. P. 565-583.

Longuett-Higgins M.S., Stewart R.W. Radiation Stress and Mass Transport in Gravity Waves with Application to "Surf Beat." *Ibid.* 1962. Vol. 13. P. 481-504.

Luck G. *Apparatus and Measuring Methods in Coastal Engineering.* Kuste. 1984. N. 41.

Lundgren H. Turbulent Currents in the Presence of Waves. *Proc. 13th Int. Coast. Eng. Conf.* Vancouver, 1972. P. 623-634.

Lushik V.G., Pavel'ev A.A., Yakubenko A.E. Transfer Equation for the Characteristics of Turbulence: Models and Results of Calculation. M. VINITI. 1988.P.3-60 (in Russian).

Lynch J.F. Theoretical Analysis of ABSS Data for HEBBLE. *Mar. Geol.* 1985. Vol. 66. P. 277-289.

Ma Y., Varadan V.K., Varadan V.V., Bedford K.W. Multifrequency Remote Acoustic Sensing of Suspended Materials in Water. *J. Acoust. Soc. Amer.* 1983. Vol. 74. P. 581-585.

Madsen O.S., Grant W. Sediment Transport in the Coastal Environment. Mass. Inst. Techn. Dep. Civ. Eng. Rep. 1976. N. 209. P. 1-120.

Makkaveev V.M. To the Theory of Turbulent Regime of Sediment Suspension. *Proc. GGI.* 1931. N. 32. P.5-27 (in Russian).

Maksimchuk V.L. Rational Application and Protection of Water Storage Basin Coasts. Kiev. Budivelnik Publ. 1981. 1981. P. 110 (in Russian).

Maksimchuk V.L. Solution of the Problem of Longshore Sediment Discharge by Means of Equations of Storm Currents. Dynamics of Wave and Circular Flows. Kiev. Naukova Dumka Publ. 1966. P. 21-25 (in Russian).

Manohar M. Mechanics of Bottom Sediments due to Wave Action. US Annu. Corp. Eng.Beach Erosion Board. Techn. Memo. 1955. N. 75. P. 1-21.

Manual on Hydrological Research in the Nearshore Sea Zone and River Mouth When Engineering Research is Being Done. N. Gidrometeoizdat publ. 1972. P. 395 (in Russian).

Manual on Calculation of Hydrological Regime Elements in the Nearshore Zone and River Mouth When Engineering Research is Being Done. M.: Gidrometeoizdat publ. 1973. P. 535 (in Russian).

Manual on Calculations of the Hydrological Parameters in the Nearshore Zone. 1973. Moscow. Hydrometeoizdat. P. 535 (in Russian).

Manual on Research and Calculation Methods of Sediment Transport and Coast Dynamics when Engineering Research is Being Done. M.: Gidrometeoizdat publ. 1975. P. 238 (in Russian).

Mantz P.A. Incipient Transport of Fine Grains and Flakes by Fluids—An Extended Shields Diagram. *J. Hydraulic Div.* 1977. Vol. 103. P. 601-615.

McDougal W.C., Hudspeth R.T. Wave Setup/Setdown and Longshore Currents on Non-Planar Beaches. *Coast. Eng.* 1982. Vol. 8. P. 103-117.

Middleton A.H.. Cahille M.L., Hsieh W.W. Edge Waves on the Sysney Coast. *J. Geophys. Res.* 1987. Vol. 92. P. 9487-9493.

Mikhailova N.A. Solid Particle Transport by Turbulent Water Flow. L.: Hydrometeoizdat publ. 1966. P. 232 (in Russian).

Miller M.C., Komar P.D. A Field Investigation of the Relationship Between Oscillation Ripple Spacing and the Near-Bottom Water Orbital Motions. *J. Sediment. Petrol.* 1980a. Vol. 50. P. 183-191.

Miller M.C., Komar P.D. Oscillation Sand Ripples Generated by Laboratory Apparatus. *Ibid.* 1980b. Vol. 50. P. 173-182.

Miller M.C., Barcilon A. Hydrodynamic Instability in the Surf Zone as a Mechanism for the Formation of Horizontal Gyres. *J. Geophys. Res.* 1978. Vol. 83. P. 4107-4116.

Mimura N., Ohtsuka Y., Watanabe A. A Laboratory Study on Two-Dimensional Beach Transformation Due to Irregular Waves. *Proc. 20th. Int. Coast. Eng. Conf.* 1986. P. 1393-1406.

Mirtzhulava T.E. Erosion of Channels and Methods of Estimation of Their Resistance. M.: Kolos Publ. 1967. P. 176 (in Russian).

Mizuguchi M. Experimental Study on Kinematics and Dynamics of Wave Breaking. *Ibid,* 1986. P. 589-603.

Mogridge G.R. Bed Form Generated by Wave Action. DME/NAE Quart. Bull. 1973. N 2. P. 1-41.

Mogridge G.R., Kamphuis J.W. Experiments on Bed Form Generation by Wave Action. *Proc. 13th Int. Coast. Eng. Conf.* Vancouver, 1972. P. 1123-1142.

Monin A.S., Yaglom A.M. Statistical Hydromechanics. M.: Nauka Publ. 1965. Part 1. P. 639 (in Russian).

Munk W.H. Surf Beats. *Trans Amer. Geophys. Union.* 1949. Vol. 30. P. 849-854.

Murina E.Y., Halfin I.S. Research of Bottom Critical Speed Under Wave Influence. *Water Resources.* 1981. N. 5. P.115-120 (in Russian).

Murray S.P. Settling Velocities and Vertical Diffusion of Particles in Turbulent Water. *J. Geophys. Res.* 1970. Vol. 75. P. 1647-1654.

Myrhaugh D. On a Theoretical Model of Rough Turbulent Wave Boundary Layers. *Ocean Eng.* 1982. Vol. 9, N. 6. P. 547-565.

Nadaoka K., Kondoh T. Laboratory Measurements of Velocity Field Structure in the Surf Zone by LDV. *Coast. Eng.* Jap. 1982. Vol.25. P.125-146.

Nadaoka K., Hino M., Koyano Y. Turbulent Flow Field Structure of Breaking Waves in the Surf Zone. *Fluid Mech.* 1989. Vol.204. P.359-387.

Nagy V.I. The Applicability of Frankl's Equations for the Investigation of Suspended Sediments. *Proc. 11th Cong. Intern. Assoc. Hydraulic Res.* Leningrad. 1965.

Nakato T. Wave-Induced Sediment Entrainment from Rippled Beds: Ph. D. Thesis. Iowa, 1974.

Nakato T., Locher F.A., Glover J.R. Kennedy J.F. Wave Entrainment of Sediment From Rippled Beds. *J. Waterways, Port, Coast. and Ocean Div.* 1977. Vol.103. P.83-100.

Natishwili O.G. Some Constructive Problems of Channel Flows Carrying Sediment and the Results of Laboratory Study of Mud Flow Movement. Ph.D. Thesis. M. 1970. P. 50 (in Russian).

Nearshore Dynamics and Coastal Processes. Tokyo. 1988. P.515.

Nguen-An-Nien. Hydrodynamical Equations for Two-Phase Flow. *Proceedings of the Hydrotechnical Meetings, Sediments Dynamics and Hydraulic Transport.* Energy Publ. 1971. N. 57. P. 73-83 (in Russian).

Nielsen P. Some Basic Concepts of Wave Sediment Transport. Progr. Rept. Inst. Hydrodyn. and Hydraulic Eng. Thechn. Univ. Denm. 1979, N 2. P. 160.

Nielsen P. Dynamics and Geometry of Wave-Generated Ripples. *J. Geophys. Res.* 1981. Vol. 86. N. 7. P.6467-6472.

Nielsen P. Field Measurements of Time-Averaged Suspended Sediment Concentrations Under Waves. *Coast. Eng.* 1984. Vol.8. P.51-72.

Nielsen P. Suspended Sediment Concentration Under Waves. *Coast. Eng.* 1986. Vol.10. P.23-31.

Nielsen P., Green M.O., Coffey F.C. Suspended Sediment Under Waves. Sydney. 1982. Dep. Geogr. Univ. Coast. Stud. Unit. Techn. Rep., N 8216.

Nikolov H.I., Kos'yan R.D. Short Term Changes of Relief and Composition of Bottom Sediments Along the Profile of Submerged Coastal Slope. *Oceanology.* S. 1990. N. 21 (in Russian).

Nikolov H.I., Pykhov N.V. Short Term Deformations of Submerged Slope Profile During Storm Interaction of Atmosphere, Hydrosphere and Lithosphere in the Coastal Zone ("Kamchia-77"). S. Bulg. Ac.Sci.Publ.1980. P.229-237 (in Russian).

Noda E.K. Wave-Induced Nearshore Circulation. *J. Geophys. Res.* 1974. Vol 79. P.4097-4106.

Nowell A.R.M., Jumars P.A., Eckman J.E. Effect of Biological Activity on the Entrainment of Marine Sediments. *Mar. Geol.* 1981. Vol.42. P.133-153.

Okayasu A., Shibayama T., Mimura N. Velocity Field Under Plunging Waves. *Proc. 20th Int. Coast. Eng. Conf.* 1987. P.660-674.

Onischenko E.L. Influence of Sedimentometer's Parameters Upon the Accuracy of Sediment Composition Analysis. *Meteorology and Hydrology.* 1987. N. 3. P.118-122 (in Russian).

Onischenko E.L., Kos'yan R.D. About the Application of Optical Method for the Determination of Suspended Sediment Concentration in Natural Basins. *Water Resources.* 1989. N 3. P.92-101 (in Russian).

Onischenko E.L., Pykhov N.V. Measuring of Suspension Fluctuation in Surf Zone. Modern Processes of Sediment Accumulation on Shelf of the World Ocean. M.: Nauka Publ. 1990. P.186-192 (in Russian).

Osborne D.P., Greenwood B. Frequency Dependent Cross-Shore Suspended Sediment Transport. 1. A Non-Barred Shoreface. *Mar. Geol.*,1992. Vol.106. P.1-24.

Osborne D.P., Greenwood B. Frequency Dependent Cross-Shore Suspended Sediment Transport, 2. A Barred Shoreface. *Mar. Geol.*,1992. Vol.106. P.25-51.

Ostendorf D.W., Madsen O.S. *An Analysis of Longshore Currents and Associated Sediment Transport in the Surf Zone.* Cambridge (Mass.). 1979. P. 88.

Ozasa H., Brampton A.H. Mathematical Modeling of Beaches Backed by Sea Walls. *Coast. Eng.* 1980. Vol.4. P.47-64.

Ozhan E. Laboratory Study of Breaker Type Effect on Longshore Sand Transport. *Proc. of Euromech 156: Mechanics of Sediment Transport.* Rotterdam. 1983. P.265-274.

Panchev S. Heavy Particle Shift in the Convective Turbulent Flow Under the Square Law of Resistance. Rep. Bulg. Ac.Sci. 1959. Vol.12. N. 4. P.51-65 (in Russian).

Papadopulos J., Ziegler C.A. Radioisotope Technique for Monitoring Sediment Concentration in Rivers and Streams. Radioisotopes. Inst. Industr. and Geophys. Symp. IAEA. SM 68/26. 1965. P.381-394.

Peregrine D.H., Swendsen I.A. Spilling Breakers, Bores, and Hydraulic Jumps. *Proc. 16th Int. Coast. Eng. Conf.* Hamburg. 1978. P.540-550.

Peregrine D.H. Wave Breaking on the Sloping Shores. *Mechanics.* 1987. N 42. P.37-72 (in Russian).

Petukhova G.A., Ivanov Y.N. Study of Bathometers for Sampling of Suspended Sediment in Water Storage Basin. *Proc.GGI.* 1965. N. 124. P.120-128 (in Russian).

Phillips O.M. *The Dynamics of the Upper Ocean.* I.: Cambridge Univ. Press. 1977. P. 366.

Phylippov A.P. Prognosis of Longshore Suspended Sediment Transport (Bulgarian coast). Ph.D. Thesis. M. I0 USSR Ac.Sci. 1988. P. 187 (in Russian).

Phylippov A.P., Kos'yan R.D., Prokof'ev V.V. Device for Study of Bottom Accumulative Form Formation and Shift on the Sea Shelf. A.C. 1236053 USSR. 1986. N 21 (in Russian).

Pils W. Mathematisches Modell fur den Feststoff transport uber Sohlformen Wasserwirtschaft. 1983. Vol.73. N. 5. P.143-148.

Podymov I.S., Kos'yan R.D. About the Possibility of Measuring of Sign-Variable Velocities in the Coastal Zone by Means of Ultrasound. Interaction of Atmosphere, Hydrosphere and Lithosphere in the Coastal Zone ("Kamchia-78"). S. Bulg. Ac.Sci.Publ. 1982. P. 139-145 (in Russian).

Pykhov N.V., Antsyferov S.M., Dachev V.D., Kos'yan R.D. Measuring of Absolute Values of Suspended Sediment Concentration During Storm. Interaction of Atmosphere, Hydrosphere and Lithosphere in the Coastal Zone ("Kamchia-78"). S. Bulg. Ac. Sci. Publ.1982. P. 146-157 (in Russian).

Pykhov N.V., Dachev V.D. Model for Calculation of Vertical Concentration Profile of the Suspended Sediment in Zone of Wave Deformation and Breaking. Interaction of Atmosphere, Hydrosphere and Lithosphere in the Coastal Zone ("Kamchia-78").S. Bulg. Ac. Sci. Publ. 1982. P.174-184 (in Russian).

Pykhov N.V., Dachev V.D., Kos'yan R.D. Changeability of Field of Suspended Sediment Concentration in Breaking Zone During Storm. Interaction of Atmosphere, Hydrosphere and Lithosphere in the Coastal Zone ("Kamchia-77"). S. Bulg. Ac. Sci. Publ. 1980. P.252-265 (in Russian).

Pykhov N.V., Dachev V.D., Kos'yan R.D., Nikolov H.I. Study of Field of Mean Concentration of Suspended Clastic Material and its Composition in the Coastal Zone. Interaction of Atmosphere, Hydrosphere and Lithosphere in the Coastal Zone ("Kamchia-77"). S. Bulg. Ac. Sci. Publ. 1980 P.238-251 (in Russian).

Pyshkin B.A., Maksimchuk V.L., Tsajts E.S. Research of Longshore Sediment Movement in Sea and Water Storage Basin. Kiev Naukova Dumka Publ. 1967. P.142 (in Russian).

Quick M.C. Wave-Induced Sand Ripples Can. *J. Civ. Eng.* 1982. Vol.9, N 2. P.285-295.

Raman H., Sethuraman V., Muralikrishna J. Velocity Measurements in Waves Using Strain Gauge Technique. *Intern. Symp. on Ocean Wave Measurements and Analysis.* ASFE. 1974. N. 4. P.817-835.

Rance R.J. *Sediment Transport by Waves Acting on a Horizontal Bed.* Wallingford. 1984. P. 25.

Reimnitz E., Ross D.A. The Sea-Sled Device for Measuring Bottom Profiles in the Surf Zone. *Mar. Geol.* 1971. N. 11. P.27-32.

Rigler J.K., Collins M.B. Initial Grain Motion Under Oscillatory Flow: A Comparison of Some Threshold Criteria. *Geo-Mar. Lett.* 1983/1984. Vol.3. P.43-48.

Rodi W. Turbulence Models and their Application in Hydraulics. Delft. 1980. P. 104.

Rossinsky K.I. Bottom Sediment Transport. *Proc. GGI.* 1968. N. 160. P.102-139 (in Russian).

Rossinsky K.I., Debolsky V.K. *River Sediments.* M.: Nauka Publ. 1980. P. 216 (in Russian).

Rossinsky K.I., Kuzmin N.A. River Channel Hydrological Basis of River Hydraulic Engineering. M. USSR Ac. Sci. Publ. 1950. P.52-97 (in Russian).

Rozhkov G.F., Ipatova Z.N., Kolobzarov O.V., Staison R.N. Fractional Grain-Size Analysis Lithology and Minerals. 1973. N. 6. P.121-135 (in Russian).

Rudowski S. Zmarszczki w strefie przybrzeza poludniowego Baltyku Acta. *Geol. pol.* 1970. Vol.20, N. 3. P.451-477.

Rybak O.K., Suprunov L.I. *Recommendations for the Calculation of Artificial Sandy Beaches.* M.: VNIITS. 1982. P. 26 (in Russian).

Rytov S.M., Vladimirskiy V.V., Galanin M.D. Sound Distribution in Dispersion Medium. *Experiment. and Theor. Phys.* 1938. Vol. 8, N. 5. P.614-622 (in Russian).

Rzhanitsyn N.A. Sediment Suspending by Turbulent Flows Under Wave Influence River Hydraulics and Hydraulic Engineering. M.: Rechizdat publ. 1952. P.28-46 (in Russian).

Salkield A.P., Le Good G.P., Soulsby R.L. An Impact Sensor for Measuring Suspended Sand Concentration. *Proc. of Electron. for Ocean Techn. Conf.* Birmingham. 1981. P.37-47.

Sallenger A.H., Holman R.A. Infragravity Waves Over a Natural Barred Profile. *J. Geophys. Res.* 1987. Vol.92. P.9531-9540.

Sallenger A.H., Holman R.A. Wave Energy Saturation on a Natural Beach of Variable Slope. *J.Geophys. Res.* 1985. Vol.90. P.11939-11944.

Sallenger A.H., Howard P.C., Fletcher C.H., Howd P.A. A System for Measuring Bottom Profile, Waves and Currents in the High-Energy Nearshore Environment. *Mar. Geol.* 1983. Vol.51. P.451-477.

Sasaki T. *Field Investigation of Nearshore Currents On a Gently Sloping Bottom.* NERC Pep. 1977. N. 3. P. 209.

Sasaki T., Horikawa K. Nearshore Current System On a Gently Sloping Bottom. *Coast. Eng. Jap.* 1975. Vol.18. P.123-142.

Sasaki T., Horikawa K., Hotta S. Nearshore Currents on a Gently Sloping Bottom. *Coast. Eng. Jap.* 1976. Vol. 18. P.123-142.

Sasaki T., Tanakashi T. Improvement of Step-Type Recording Wave Gauge with the Application of Microprocessor. Rep. Port and Harbour Res. Inst. 1983. Vol.22. N. 3. P.57-82.

Sato S., Uehara H., Watanabe A. Numerical Simulation of the Oscillatory Boundary Layer Flow Over Ripples by a K-E Turbulent Model. *Coast. Eng. Jap.* 1986. Vol.29. P.65-78.

Sawamoto M. Flow Field over Beds Induced by Wave Action. *Proc. 3rd Intern. Symp. on Stochastic Hydraulic,* Tokyo. 1980. P.621-630.

Schemer E.W., Schubel I.R. A Near Bottom Suspended Sediment Sampling System for Studies of Resuspension. *Limnol. and Oceanogr.* 1970. Vol.5. N. 4. P.644-646.

Schmidt W. *Der Messenaustausch in Probleme der Kosmischen Physik*. Hamburg. 1925. Bd. 7. P.63.

Sert M. Transition in Oscillatory Boundary Layers. Euromech 156: Mechanics of Sediment Transport. Istanbul, 1982. P.255-263.

Shadrin I.F. *Currents in the Coastal Zone of the Nontidal Sea*. M.: Nauka Publ. 1972. P. 128 (in Russian).

Shapovalov P.B. *Silting of Zhdanov Channel*. Rostov-on-Don. 1956. P. 87. (in Russian).

Shi N.C., Larsen L.H., Downing J.P. Predicting Suspended Sediment Concentration on Continental Shelves. *Mar. Geol*. 1985. Vol.62. P.255-275.

Shibayama T., Horikawa K. Sediment Suspension due to Breaking Waves. *Coast. Eng. Jap*. 1982. Vol.25. P.163-176.

Shibayama T. Sediment Transport Mechanism and Two-Dimensional Beach Transformation due to Waves. Ph.D. Thesis. Tokyo. 1984. P. 159.

Shields A. *Anwendung der Achulikeitsmecanik und der Turbulentforschung auf die Geschiebebewegung*. Berlin. 1936. P. 20.

Shifrin K.S. Optical Study of Cloud Particles. Investigations of Clouds, Precipitation and Thunderstorm Electricity. L.Gidrometeoizdat publ. 1957. P.19-24 (in Russian).

Shifrin K.S., Golikov V.I. Measuring of Microstructure by Narrow Angle Method. *Proc. State hydrol. observatory*. 1964. N 152. P.121-137 (in Russian).

Shimizu T., Saito S., Maruyama T. et al. Modelling of Onshore-Offshore Sand Transport Rate Distribution Based on Large Wave Flume Experiments. Cent. Res. Inst., Electric Power Industry, Civ. Eng. Lab. Rep. N 384028. 1985. P. 60.

Shinohara K., Tsubaki T., Joshitaka M., Agemori C. Sand Transport Along Sandy Beach by Wave Action. *Coast. Eng. Jap*. 1958. Vol.1. P.111-131.

Shore Protection Manual 1984. 4th.ed., 2 Vols. US Army Engineer Waterways Experiment Station, Coast. Eng.Res.Center, US Government Printing Office, Washington, DC.

Shraiber A.A. About Diffusion of Heavy Particle in Turbulent Flow. *Heat Physics and Heat Engineering*. 1973b. Vol.25. P.110-113 (in Russian).

Shraiber A.A. About Fluctuated Movement of Small Particle of Discrete Phase in Turbulent Flows. *Heat Physics and Heat Engineering*. Kiev. 1973a. Vol.25. P.73-79 (in Russian).

Shtelzer K. Calculation of the Coarse-Grained Sediment Motion. *Water Resources*. 1984. N. 1. P. 3-9 (in Russian).

Shulyak B.A. *Wave Physics on the Surface of Loose Medium and Fluid*. M.: Nauka publ. 1971. P. 399 (in Russian).

Shwartsman A.Y. Investigation and Calculation of the Turbidity in the Nearshore Zone of Water Storage Basin. *Proc. GGI*. 1965. N. 124. P.17-31 (in Russian).

Sinelschikov V.S. Distribution of Suspension Concentration in Turbulent Two-Phase Flows. *Proc. USSR Ac. Sci. OTN. Mechanics and engineering*. 1963. N. 1. P.150-152 (in Russian).

Singamsetti S.R. Diffusion of Sediment in a Submerged Jet. *J. Hydraulic Div*. 1966. Vol.92, N 2. P.153-169.

Skafel M.C., Krishnappan B.G. Suspended Sediment Distribution in Wave Field. *J. Waterways, Port, Coast. and Ocean Eng*. 1984. Vol110. P.215-230.

Sleath J.F.A. A Contribution to the Study of Vortex Ripples. *J. Hydraulic Res.* 1975. Vol.13. N. 3. P.315-328.

Sleath J.F.A. On Rolling Grain Ripples. *J. Hydraulic Res.* 1976. Vol.14. N 1. P.69-81.

Sleath J.F.A. The Suspension of Sand by Waves. *J. Hydraulic Res.* 1984. Vol.20. P.439-452.

Sleath J.F.A. *Sea Bed Mechanics.* N.Y.: Wiley. 1984. N 4. P. 355.

Sleath J.F.A. Turbulent Oscillatory Flow Over Rough Beds. *J. Fluid. Mech.* 1987. Vol.182. P.369-409.

Sleath J.F.A. Ellis A.S. Ripple Geometry on Oscillatory Flow. Univ. of Cambridge, Department of Eng., 1978, Rep. NA/Hydraulics/TR 2. P. 15.

Slezkin N.A. Differential Equations of Pulp Movement. Rep. USSR Ac.Sci. 1952. Vol. 86, N 2. P.235-237 (in Russian).

Smith J.D. Modelling of Sediment Transport on Continental Shelves. *The Sea.* N.Y.: Wiley. 1977. Vol.6. P.539-577.

Smith J.D., Kraus N.C. Longshore Currents Based on Power Law Wave Decay. *Proc. Coast. Hydrodyn.*, ASCE. 1987. P.155-169.

Smith J.D., McLean S.R. Spatially Averaged Flow Over a Wavy Surface. *J. Geophys. Res.* 1977. Vol.82. P.1735-1746.

Smutek R. Pohybpenve Castice v Turbulentum Proudu. Vodohospod. Cas. 1974. Roz.22, Z3. P.254-269.

Sokolov R.N., Kudryavitsky F.A., Petrov G.D. Underwater laser instrument for measuring the spectrum of suspended particle size in the sea. *Proc. USSR Ac. Sci. Physics of atmosphere and ocean.* 1971. Vol. 7. N. 9. P.1015-1018 (in Russian).

Solov'ev N.Y. Improvement and Testing of a Coarse Sediment Recorder. *Proc. GGI.* 1967. N 141. P.58-78 (in Russian).

Sonu C.J. Field Observation of Circulation and Meandering Currents. *J. Geophys. Res.* 1972. Vol.77. P.783-805.

Soo S.L. Statistical Properties of Momentum Transfer in Two Phase Flow. *Chem. Eng. Sci.* 1956. Vol.5, N 2. P.57-63.

Sorokin V.M., Dimitrov P.S. Rates of Sediment Accumulation on the Continental Terrace in the Late Quaternary Period. Geologogeophysical Study of Bulgarian Section of the Black Sea. S. Bulg. Ac. Sci. Publ. 1980. P.238-245 (in Russian).

Soulsby R.L. The Bottom Boundary Layer of the Shelf Seas. *Physical Oceanography of Coastal and Shelf Seas.* Amsterdam: Elsevier. 1983. P.189-266.

Soulsby R.L., Salkield A.P., Haine R.H., Wainwright B. Observations of the Turbulent Fluxes of Suspended Sand Near the Sea Bed. *Transport of Suspended Solids in Open Channels.* Rotterdam: Balkema. 1986. P.295-324.

Soulsby R.L., Salkield A.P., Le Good G.P. Measurements of the Turbulence Characteristics of Sand Suspended by a Tidal Current. *Cont. Shelf. Res.* 1984. Vol.3. P.439-454.

Soulsby R.L., Atkins R., Salkield A.P. Observations of the Turbulent Structure of a Suspension of Sand in a Tidal Current. Mechanics of Sediment Transport in Fluvial and Marine Environments. Genova. 1987. P.88-91.

Soy S. *Hydrodynamics of Multi-Phase Systems.* Moscow: Mir Publ. 1971. P. 536 (in Russian).

Speransky N.S., Leont'ev I.O. Technique of Transfer Rate Determination in the Wave Flow. *Oceanology*, 1979. Vol.19. N 3. P.514-519 (in Russian).

Staub C., Swendsen I.A., Johnsson I.G. Measurements of the Instantaneous Sediment Suspension in Oscillatory Flow. Progr. Inst. Hydrodyn. and Hydraulic Eng. Techn. Univ. Denm. 1983. P. 41-49.

Sternberg R.W., Larsen L.H. Threshold of Sediment Movement by Open Waves. *Deep-Sea Res.* 1975. Vol.22. P.299-309.

Sternberg R.W., Shi N.C., Downing J.P. Field Investigation of Suspended Sediment Transport in the Nearshore Zone. *Proc. 19th Int. Coast. Eng. Conf.* Houston. 1984. P.1782-1798.

Stive M.J.F., Wind H.G. A Study of Ration Stress and Set-Up in the Nearshore Region. *Coast. Eng.* 1982. Vol.6. P.1-25.

Stive M.J., Wind H.G. Cross Shore Mean Flow in the Surf Zone. 1986. *Coast. Eng.* Vol.10. P.325-340.

Sudolsky A.S. Longshore Currents in the Shallow Water of the Water Storage Basin. *Proc. GGI*, 1963. N 106. P.174-181 (in Russian).

Sudolsky A.S., Yaroslavtsev N.A. Main Feature and Peculiarities of Longshore Currents and Sediment Transport in the Shallow Water Storage Basins. *Proc. of IV ALL union Hydrol. Congress. M.*: Gidrometeoizdat publ. 1976. Vol. 10. P. 301-314 (in Russian).

Summer B.M. Recent Developments on the Mechanics of Sediment Suspension. *Euromech 192*: Transport of Suspended Solids in Open Channels. Rotterdam: Balkema, 1986. P.3-13.

Summer B.M., Deigaard R. Particle Motion Near the Bottom in a Turbulent Flow in an Open Channel. Pt. 2. *J. Fluid Mech.* 1981. Vol.109. P.311.

Summer B.M., Oguz B. Particle Motion Near the Bottom in a Turbulent Flow in an Open Channel. *J. Fluid Mech.* 1978. Vol.86. P.109.

Summer B.M., Jensen B.L., Fredsøe J. Turbulence in Oscillatory Boundary Layers. *Advances in turbulence.* Ed. G. Comte-Bellot, Y. Mathion. Berlin: Springer, 1987. P.556-567.

Sunamura T. A Laboratory Study of Offshore Transport and a Model for Eroding Beaches. *Proc. 17th Int. Conf. Coast. Eng.* Sydney. 1981. P.1051-1070.

Sunamura T. Production of Onshore-Offshore Sediment Transport Rate in the Surf Zone. Including Swash Zone. *Proc. 31th Jap. Conf. Coast. Eng.* Tokyo, 1984. P.316-320.

Sunamura T., Bando K., Horikawa K. An Experimental Study of Sand Transport Mechanism and Rate over Asymmetrical Sand Ripples. *Proc. 25th Jap. Conf. Coast. Eng.* Tokyo. 1978. P.250-254.

Suy Tsy-fyn. Application of the Sea Waves Theory for the Calculation in the Coastal Zone. *Proc. P.P. Shirshov Institute* USSR Ac. Sci. 1965. Vol.76. P.189-224 (in Russian).

Sverdrup H.V., Munk W.H. Wind, Sea and Swell: Theory of Relations for Forecasting. Hydrogr. off. pub. 1947. N 601. P.1-44.

Swart D.H. Computation of Longshore Transport. Delft Hydraulic Lab. Rep. Invest. 1976. N 968, pt 1. P.1-112.

Swendsen I.A. Analysis of Surf Zone Turbulence. *J.Geophys. Res.* 1987. Vol.92. P.5115-5124.

Swendsen I.A. Mass Flux and Undertow in a Surf Zone. *Coast. Eng.* 1984. Vol.8. P.347-365.

Swendsen I.A. Mass Flux and Undertow in a Surf Zone, by I. A. Swendsen—Reply. *Coast. Eng.* 1986. Vol.10. P.299-307.

Swendsen I.A., Schaffer H.A., Hansen J.B. The Interaction Between the Undertow and the Boundary Layer Flow on a Beach. *J. Geophys. Res.* 1987. Vol.92. P.11845-11856.

Symonds G., Bowen A.J. Two-Dimensional Surf Beat: Long Wave Generation by a Time Varying Break-Point. *J. Geophys. Res.* 1982. Vol.87. P.492-498.

Symonds G., Bowen A.J. Interaction of Nearshore Bars with Incoming Wave Groups. *Ibid.* 1984. Vol.89. P.1953-1959.

Taggart W.C., Yermoli C.A., Montes S., Ippen A.T. Effect of Sediment Size and Gradation of Concentration Profiles for Turbulent Flow. M.I.T. Lab. Water Resour. and Hydrodyn. Rep. 1972. N 152. P. 154.

Tanner W.H. Numerical Estimates of Ancient Waves, Water Depth and Fetch. *Sedimentology.* 1971. Vol.103, N 4. P.439-442.

Tchen Chan-mou. Mean Value and Correlation Problems Connected With the Motion of Small Particles Suspended in a Turbulent Fluid. Ph.D. Thesis. The Haag. 1947.

Thornton E.B. Energetics of Breaking Waves Within the Surf Zone. *J. Geophys. Res.* 1979. Vol. 84. P.4931-4939.

Thornton E.B. Variation of Longshore Currents Across the Surf Zone. *Proc. 12th Int. Conf. Coast. Eng.* 1970. P.291-308.

Thornton E.B., Guza R.T. Surf Zone Longshore Currents and Random Waves: Filed Data and Models *J. Phys. Oceanogr.* 1986. Vol. 16. P.1165-1178.

Trowbridge J., Madsen O.S. Turbulent Wave Boundary Layers, 1. Model Formulation and First-Order Solution. *J. Geophys. Res.* 1984a. Vol.89. N 5. P.7989-7997.

Trowbridge J., Madsen O.S. Turbulent Wave Boundary Layers, 2. Second-Order Theory and Mass Transport. *J. Geophys. Res.* 1984b. P.7998-8005.

Tucker M.J. Surf Beats Sea Waves of 1 to 5 min. Period. *Proc. Roy. Soc. London A.* 1950. N 202. P.565-573.

Tunstall E.B., Inman D.L. Vortex Generation by Oscillatory Flow over Rippled Surfaces. *J. Geophys. Res.* 1975. Vol. 80. P. 3475-3485.

Tsajts E.S., Homitsky V.V. Research of Longshore Sediment Run-Off When the Coastal Slope is Changing. *Hydromechanics.* 1978. N 38. P.66-70 (in Russian).

Urick R.J. The Absorption of Sound in Suspensions of Irregular Particles. *Acoust. Soc. Amer.* 1948. Vol.3. P.238-289.

Ursell F. Edge Waves on a Sloping Beach. *Proc. Roy. Soc. London A.* 1952. Vol.214. P.79-97.

Van De Graaf J., Van Overeem J. Evaluation of Sediment Transport Formulae in Coastal Engineering Practice. *Coast. Eng.* 1979. Vol.3. P.1-32.

Van Rijn L.C. Mathematical Models for Sediment Concentration Profiles in Steady Flow. *Transport of Suspended Solids in Open Channels.* Rotterdam: Balkema, 1986. P.49-68.

Vanoni V., Nomicos G.H. Resistance Properties of Sediment-Laiden Streams. *Trans. Amer. Soc. Civ. Eng.* 1960. Vol.125. P.1140-1148.

Vanoni V. Transportation of Suspended Sediment by Water. *Trans. Amer. Soc. Civ. Eng.* 1946. Vol.111. P.67-134.

Varadan V.V., Ma Y., Varadan V.K. Theoretical Analysis of the Acoustic Response of Suspended Sediment for HEBBLE. *Mar. Geol.* 1985. Vol.66. N 1/4. P.267-276.

Velikanov M.A. Transport of Suspended Sediment by Turbulent Flow. 1944. *Proc. USSR Ac.Sci. OTN.*1944. N 3. P.189-208 (in Russian).

Velikanov M.A. *Dynamics of Channel Flows.* M.: Gostechizdat. 1955. Vol.2. P. 323 (in Russian).

Vershinskiy N.V. About Measuring of Oscillating Processes in the Sea. *Meteorology and Hydrology.* 1951. N 6. P.51-54 (in Russian).

Vershinskiy N.V. To the Problem of Force Factors in the Coastal Zone. Reports of USSR Ac.Sci. 1952 Vol. 87, N 5, P.723-726 (in Russian).

Vincent C.E., Young R.A., Swift D.J.P. On the Relationship Between Bedload and Suspended Load Transport on the Inner Shelf. *J. Geoph. Res.* 1982. Vol.87. P.4163-4170.

Vinogradova V.I. Experimental Research of Velocity Fields and Concentration of Sediment Carrying Flows. Ph.D. Thesis.1962. P. 20 (in Russian).

Vladimirskiy V.V., Galanin M.D. Ultrasound Absorption in Aqueous Emulsion of Mercury. *J. Experim. and Theor. Physics.* 1939. Vol. 9, N 2. P.233-236 (in Russian).

Voitsekhovich O.V. Field Measuring and Estimation of Some Energy Methods of Sediment Discharge Calculation in the Nearshore Zone. *Problems of Sediment Transport in the Coastal Zone.* Tbilisi. 1983. P.6-8 (in Russian).

Voitsekhovich O.V. Field Study of Storm Current Sand of Longshore Sediment Transport on the Black Sea Northern-Western Coast. Ph.D. Thesis. 1986a. P. 20 (in Russian).

Voitsekhovich O.V. Longshore Sediment Transport—Summarized Dependencies and Field Data. *Water Resources.* 1986b. N 5. P.108-116 (in Russian).

Vongvisessomjai S. Oscillatory Ripple Geometry. *J. Hydraulic Eng.* 1984. Vol.110. P.247-266.

Vyatr. Climatic Book of the Bulgaria. 1982. Vol. 4 (in Bulgarian).

Wang H., Liang S.S. Mechanics of Suspended Sediment in Random Waves. *J. Geophys. Res.* 1975. Vol.80, N 24. P.3488-3494.

Watanabe A. Numerical Models of Nearshore Currents and Beach Deformation *Coast. Eng.* Jap. 1982. Vol. 25. P.147-161.

Watanabe A., Rino Y., Horikawa K. Beach Profiles and Onshore-Offshore Sediment Transport Proc. 17th *Int. Conf. Coast. Eng.* Sydney. 1981. P.1106-1121.

Watts G.M. Development and Field Test of a Sampler for Suspended Sediment in Wave Action. *Beach Eros. Board Techn. Mem.* 1953. Vol.34. P.1-41.

Watts G.M. Field Investigation of Suspended Sediment in Surf Zones. *Proc. 4th Int. Conf. Coast. Eng.* Chicago, 1954. P.81-95.

Wiberg P.W., Smith J.D. A Comparison of Field Data and Theoretical Models for Wave-Current Interactions at the Bed on the Continental Shelf. *Conf. Shelf Res.* 1983. Vol.2. P.147-162.

Williams A.J. BASS: An Acoustic Current Meter Array for Benthic Flow-Field Measurements. *Mar. Geol.* 1985. Vol.66. N 1/4. P.345-356.

Wind H.G., Vreugdenhil C.B. Rip Current Generation Near Structures *J. Fluid. Mech.* 1986. Vol.171. P.459-476.

Worthy A.L. Wind-Generated, High-Frequency Edge Waves. *J. Mar. and Fresh-Water Res.* 1984. Vol.35. P.1-7.

Wu C.S., Liu P.L.F. Finite Element Modeling of Nonlinear Coastal Currents. *J. Waterways Port, Coast. and Ocean Eng.* 1985. Vol.111. P.417-432.

Wu C.S., Thornton E.B., Guza R.T. Waves and Longshore Currents: Comparison of a Numerical Model with Field Data. *J. Geophys. Res.* 1985. Vol.90, N 3. P.4951-4958.

Yalin M.S. On the Determination of Ripple Length. *J. Hydraulic Div.* 1977. Vol.103, N 4. P.439-442.

Yalin M.S., Karahan E. On the Geometry of Ripples due to Waves. *Proc. 16th. Int. Conf. Coast. Eng.* Hamburg. 1978. Vol.2. P.1776-1786.

Yalin M.S., Karahan E. Inception of Sediment Transport. *J. Hydraulic Div.* 1979. Vol.105. P.1433-1443.

Yalin M.S., Russel R.C.H. Similarity in Sediment Transport due to Waves. *Proc. 8th Int. Conf. Coast. Eng.* Mexico. 1962. P.151-171.

Yamashita T., Sawamoto Y., Yokoyama H. Experimental Study on the Sand Transport Rate and the Mechanism of Sand Movement due to Waves. *Proc. 31th Jap. Conf. Coast. Eng.* 1984. P.281-285.

Yang W. Surf Zone Properties and On-Offshore Sediment Transport. Ph.D. Thesis. Delaware. 1981.

Yasiewicz R. Badanie Rozkladu Unosin w Rzekach. *Gosp. Wodna.* 1973. Vol.33. N 11/12.

Young R.A., Merrill Y., Proni Y.R., Clarke T.L. Acoustic Profiling of Suspended Sediment in the Marine Boundary Layer *Geophys. Res. Lett.* 1982. N 9. P.175-178.

Yurkevich M.G. Study of Relief Deformation in the Nearshore Zone. *Investigation of Relief Dynamics of the Sea Coast.* M. Nauka Publ. 1979. P.50-64 (in Russian).

Yurkevich M.G. Kirlis V.I., Dolotov Y.S. et al. Peculiarities of Relief Changes and Differentiation of Sediments on the Beach Surface under Different Wave Regime. Interaction of Atmosphere, Hydrosphere and Lithosphere in the Coastal Zone ("Kamchia-78"). S. Bulg. Ac. Sci. publ. 1982. P.185-204 (in Russian).

Zagustin A., Zagustin K. Mechanics of Turbulent Flow in Sediment-Laden Streams Bol. Lab. Hydraulic Univ. cent. Venez. 1969. N 2. P.114-124.

Zenkovich V.P. *The Principles of Sea Shore Evolution Doctrine.* M.: USSR Ac. Sci. Publ. 1962. P. 710 (in Russian).

Zheleznyak M.I. About the Structure of Near Bottom Turbulent Wave Boundary Layer. *Hydromechanics.* 1988. N 58. P.1-8 (in Russian).

Appendix

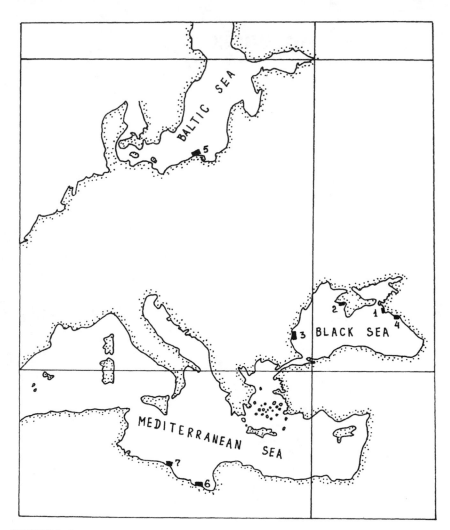

FIGURE A-1   Sketch-map of location of field measurements: 1: Anapa; 2: Donuzlav; 3: Kamchia; 4: Chernomor; 5: Lubiatowo; 6: Sirt; 7: Homs.

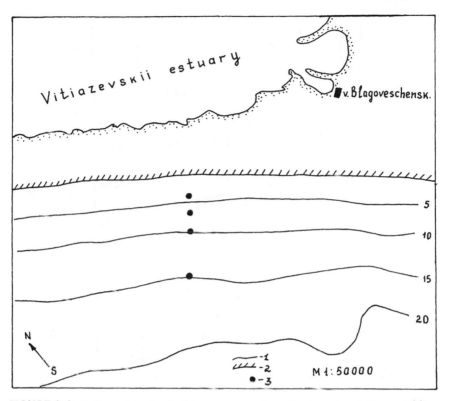

FIGURE A-2   Bathymetric chart of Anapa measuring site. 1: contours; 2: coastal line; 3: points of measuring post location.

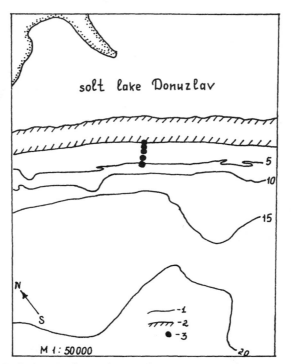

FIGURE A-3 Bathymetric chart of Donuzlav measuring site. 1: contours; 2: coastal line; 3: points of measuring post location.

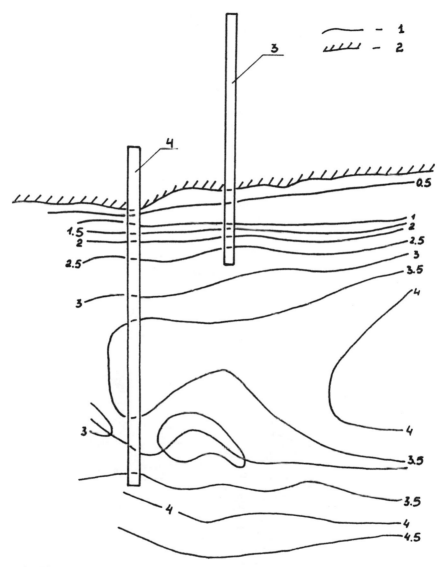

FIGURE A-4 Bathymetric chart of Kamchia measuring site. 1: contours; 2: coastal line; 3: old testing bridge; 4: new testing bridge.

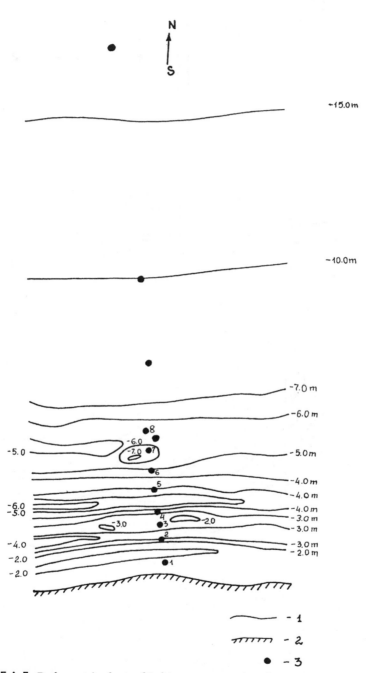

FIGURE A-5 Bathymetric chart of Lubiatowo measuring site. 1: contours; 2: coastal line; 3: points of measuring post location.

Index

DATE DUE

HIGHSMITH

Coastal Processes in Tideless Seas comprehensively addresses the complex physics affecting the movement and distribution of noncohesive sediments. This long-awaited translation and update of the 1991 original describes the results of field research of sand sediment dynamics in the coastal zones of the Black, Baltic, and Mediterranean Seas.

Conveniently divided into three parts—modern concepts of nearshore hydrodynamics and elementary coastal processes; field research; and sediment transport—*Coastal Processes* is an easy-to-use reference guide for students, engineers, and researchers interested in the problems of sediment transport in the coastal zone.

Chapters include:

- Nearshore Hydrodynamics
- Elementary Hydrodynamic Transport Processes
- Measuring Methods
- Characterization of Testing Sites, Description of Experiments, and Observation Data
- Study of Bed Microforms in the Nearshore Zone
- Suspended Sediment in the Nearshore Zone
- Longshore Sediment Transport
- Cross-Shore Sediment Transport and Variability of the Underwater Slope Profile

About the Authors

Ruben D. Kos'yan is Head of the Nearshore Department, Southern Branch, P.P. Shirshov Institute of Oceanology, Russian Academy of Sciences. He has published countless papers and several monographs. Dr. Kos'yan has won several awards, including First Prize among Young Scientists of the U.S.S.R. and the Big Silver Medal Award from the All-Union Exhibition of Achievement in the U.S.S.R. National Economy. He is a Russian National Coordinator of the researches in the Black Sea.

Nikolay Pykhov is Leading Scientist, P.P. Shirshov Institute of Oceanology, Russian Academy of Sciences, Moscow. Dr. Pykhov has authored numerous scientific articles and monographs.

Billy L. Edge is W. H. Bauer Professor of Dredging Engineering and Head, Coastal and Ocean Engineering Division, Department of Civil Engineering, Texas A&M University. He serves as Director, American Shore and Beach Preservation Association; Secretary, Coastal Engineering Research Council of ASCE; a member of the Marine Board of the National Research Council; and a member of the Coastal Engineering Research Board of the U.S. Army Corps of Engineers. Dr. Edge is also Editor of the Proceedings of the International Conference on Coastal Engineering. He is the recipient of many awards, including ASCE's Arthur M. Wellington Prize.

ISBN 0-7844-0018-0

9 780784 400180